SOURCES AND METHODS IN GEOGRAPHY

Urban Data Sources

SOURCES AND METHODS IN GEOGRAPHY

Series editors

M.A. Morgan PhD
Department of Geography, University of Bristol

D.J. Briggs PhD
Department of Geography, University of Sheffield

SOURCES AND METHODS
IN GEOGRAPHY

Urban Data Sources

John R. Short MA, PhD
Department of Geography,
University of Reading

BUTTERWORTHS
London - Boston
Sydney - Wellington - Durban - Toronto

The Butterworth Group

UK London
Butterworth & Co (Publishers) Ltd
88 Kingsway, WC2B 6AB

Australia Sydney
Butterworths Pty Ltd
586 Pacific Highway, Chatswood, NSW 2067
also at Melbourne, Brisbane, Adelaide and Perth

South Africa Durban
Butterworth & Co (South Africa) (Pty) Ltd
152–154 Gale Street

New Zealand Wellington
Butterworths of New Zealand Ltd
T & W Young Building, 77–85 Customhouse Quay,
1, CPO Box 472

Canada Toronto
Butterworth & Co (Canada) Ltd
2265 Midland Avenue
Scarborough, Ontario M1P 4S1

USA Boston
Butterworth (Publishers) Inc
10 Tower Office Park, Woburn, Mass. 01801

First published 1980

ISBN 0 408 10640 9

Typeset by Scribe Design, Gillingham, Kent
Printed and bound by Fakenham Press Ltd, Fakenham, Norfolk

British Library Cataloguing in Publication Data

Short, John R
Urban data sources. – (Sources and methods in geography).
1. Cities and towns. 2. Geography – Methodology
I. Title II. Series
910.09'1732 GF125 79-42888

ISBN 0–408–10640–9

FOREWORD

During recent years, geography has been undergoing considerable change. There have been many facets to this change, but one underlying theme is the adoption of a more rigorous approach to geographical enquiry, wherever this is appropriate. It has been reflected in numerous ways: in the greater emphasis which is placed upon quantitative and statistical methods of data collection and handling; in the attention given to the study of process as opposed to the description of form in human as well as in physical geography; and in the use of an inductive rather than deductive philosophy of learning.

What this means in practical terms is that the student and teacher of geography need to be acquainted with a wide range of methods. The student, both at school and in higher education, is increasingly becoming involved in projects or classwork which include some form of individual and original research. To be equipped for this type of study he needs to be aware of the sources from which he can obtain data, the techniques he can use to collect this information and the approaches he can take to analyse it. The teacher similarly requires a pool of empirical material on which he can draw as a source of class exercises. Both must be able to tackle geographical problems in a logical and scientific fashion, to construct appropriate explanatory hypotheses, and test these hypotheses in an objective and rational manner.

The aims of this series of books are therefore to introduce a range of sources which provide information for project and classwork, and to outline some of the methods by which this material can be analysed. The main concern is with relatively simple approaches rather than more sophisticated methods.

The reader will be expected to have a basic grounding in geography, and in some of the books a working knowledge of mathematical methods is useful. The level of detail and exposition, however, is intended to make it possible for the student, with little further reading, to gain a basic understanding of the selected themes. Consequently the series will be of particular interest and use to students and teachers involved in courses in which practical and project work figure as major components. At the same time students in higher education will find the books an invaluable guide to geographical methods.

M.A. Morgan   D.J. Briggs

ACKNOWLEDGEMENTS

The production of this book would have been impossible without the kind help of a number of people. In particular, Dr. Michael Morgan suggested the topic, read my handwritten drafts — a truly Herculean task — and compiled the collages in Chapter 7; Sheila Dance and Brian Rogers drew most of the diagrams and responded marvellously to my incessant demands and Martin Boddy showed me, in a series of extra-mural lectures with Keith Bassett, how certain data sources could be used to identify patterns of ownership.

I am also grateful to the following who have kindly given me permission to reproduce the cited items:

Mr. John D. McLachlan (Dun & Bradstreet Ltd.) for permission to reproduce the material in ***Figures 6.12*** and ***6.13***;

Mr. R.J.E. Dangerfield (Kompass Publishers Ltd.) for permission to reproduce the material in ***Figure 6.11***;

Mr. Paul Surridge (Charles E. Goad Ltd.) for permission to reproduce ***Figure 6.4***;

Mr. T.J. Ulrick (Kelly's Directories Ltd.) for permission to use the extracts in ***Figure 6.1***;

The Director, Office of Population Censuses and Surveys for permission to reproduce ***Tables 3.1, 3.2*** and ***3.3***.

PREFACE

Three strands, or to be more accurate, two strands and a loose end, constitute the contemporary literature of interest to geographers: a focus on the concepts and theories used to explain and understand the world around us; a concern with the techniques which can be used to test the hypotheses and ideas generated by these theories and; an interest in the data sources which allow these techniques to be used. Unfortunately, these three strands are generally represented in separate texts. General textbooks seek to present and sometimes evaluate the body of research findings from one field of enquiry. Such books are marked by an interest in the results of previous work rather than how such results were obtained. There is now a plethora of textbooks concerned with the elaboration of the techniques of research — the most common being the statistical technique textbook. These volumes are noted for their interest in the methods of research rather than the theoretical significance of the results. Finally, there are those tomes, small in number, which seek to provide a comprehensive introduction to data sources. Apart from the notable exception of the present series, such texts are simply concerned with highlighting the range of available data sources.

The aim of this book is to weave these three strands within the covers of a single text. Data sources will be discussed with reference to the way in which they can be used to test hypotheses and concepts concerning the structure of cities, while the elaboration of the different concepts of urban structure and social processes will be presented in a context which seeks to show how these concepts can be tested by the use of certain techniques on specific kind of data.

The form of the book is predicated upon that old adage, 'listen and know for a day, see and know for a week, do and know for a lifetime'. The best form of education is active, with both the teacher and the student embarking on a voyage of discovery. The voyage is more interesting if the ship's company consists of active crew members rather than passive passengers. This book should be seen as a manual for the crew, not a diversion for the passengers.

J.R.S.

CONTENTS

ILLUSTRATIONS

TABLES

CHAPTER 1 INTRODUCTION

1.1 AIMS AND OBJECTIVES

The aim of this book is to introduce a range of data sources which can be used to test hypotheses and theories concerning the structure of cities. In the following chapters data sets and hypotheses related to different aspects of towns and cities will be examined. As the title suggests, the main focus of this book will be on the range of available data sources and the consequent limitations and opportunities which these impose and confer on our understanding of urban areas. Before examining these it is important to note that data, by which we mean facts and figures, are used by geographers and other social scientists in conjunction with hypotheses and statistical techniques. The relationships between these three elements are portrayed in ***Figure 1.1***. In this chapter we will examine more closely each of these elements and the nature of their inter-relationships, so that we can better understand some of the problems presented by the material that follows. The discussion of hypotheses will be pitched at a general level, while the discussion of data and techniques will highlight specific problems which the reader will encounter in the use and analysis of urban data sources.

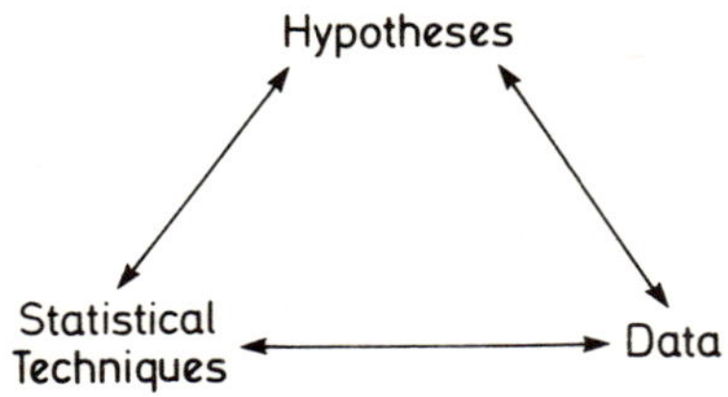

Figure 1.1 Elements of research

1.2 HYPOTHESES

Hypotheses can be defined as questions or propositions about the world around us. They can range from the fundamental, 'Is the world flat?', to the general, 'What is the structure of land uses in a city?', to the specific, 'Do persons of similar occupation live close together in cities?' To a large extent the questions which we ask are guided by the results of previous studies. For example, on the basis of an analysis of Chicago, the sociologist Burgess argued that persons in low income occupations were located close to the city centre while persons in higher income occupations lived further away from the city centre (***see Figure 1.2***). We can use this theory to form a hypothesis, 'Is the distribution of occupational categories in British cities related to distance from the city centre?'

The importance of an initial hypothesis lies in the fact that it provides the perspective for choosing the appropriate data and techniques. There are literally millions of pieces of data and hundreds of techniques which can be used to study various aspects of cities. In the murky depths of a sea of data and techniques, hypotheses provide us with a powerful beam to select specific data and appropriate techniques. If, for example, we wished to test the hypothesis that persons of different occupations live at different distances from the city centre, then we would try to obtain information on the location of these occupational categories and plot this information with reference to distance from the city centre. With an initial hypothesis we know exactly the type of data and the technique which we require.

1.2.1 Hypotheses and data

The relationship between hypotheses and data is not simply one of hypotheses guiding the selection of data. As ***Figure 1.1*** suggests, there is an interaction between the two elements. To an extent the availability of data guides our initial hypotheses. If we have a hypothesis but no data on which to verify this hypothesis, then our proposition remains in the realm of speculation. This is not to suggest that hypotheses incapable of verification are worthless — speculation is a laudable activity — but merely to note that unverifiable hypotheses lack analytical power. In certain circumstances the hypothesis which we want to verify may oblige us to generate the appropriate data. If we wish to discover the present relationship between occupation and distance, then we would need to generate contemporary data, since the latest official statistics refer only to 1971. There is, however, a limit, formed by the constraints of time and money, to the extent to which we can generate data. It is highly unlikely that one person, or indeed a whole sixth form in a school, could collect this data for all major cities in Britain. A more feasible proposition would be to test the relationship for one city. In this case the hypothesis would be, 'What is the distribution of occupational categories in (e.g.) Birmingham?'

1.2.2 Hypotheses and techniques

The initial hypothesis guides the selection of data and to an extent is guided by the availability of data; similarly with the techniques which we use to manipulate the data. The appropriate technique is the one which yields a definite answer to our initial question (this is termed verification) and in turn the acceptance of the answer depends on the suitability and rigour of the technique. To continue with our previous example, if information is obtained on the residence and occupation of individuals and we test the relationship between occupation and distance from the city centre by enumerating the numbers in

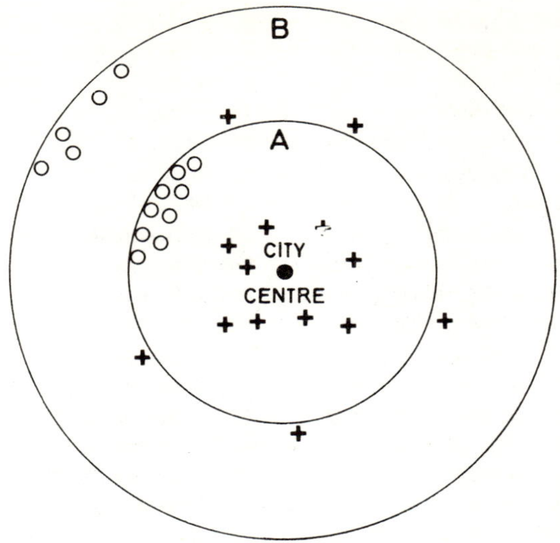

	Zone A	Zone B
Nos. of bricklayers	9	5
Nos. of surgeons	9	5

Figure 1.2 Location of two occupation groups

each occupation for two distance bands, less than 5 km and more than 5 km from the city centre, then this technique lacks the precision necessary to answer the hypothesis definitely. Think of the reasons why this technique is inappropriate. As the hypothetical situation in *Figure 1.2* highlights, if we merely use the totals for each zone then it appears that there is no difference in the distances from the city centre for the location of bricklayers and surgeons, yet as the diagram illustrates, this masks the considerable variation within the distance zones.

1.2.3 Testing of hypotheses

Hypotheses provide us with a guide for selecting types of data, while techniques allow us to manipulate the data in order to test, i.e. verify, the hypotheses. After we have raised the initial hypothesis, found the necessary data and used the appropriate technique to manipulate the data, we are then in a position to either accept or reject the hypothesis. If the hypothesis is not confirmed, for example if we find that there is no relationship between occupation and distance of residence from the city centre, we may need to re-examine the data and the technique. Were they appropriate to the hypothesis? Eventually we may have to accept that the hypothesis is incorrect. One scientist (T.H. Huxley), obviously speaking from bitter experience, once remarked that there was no greater tragedy in science than the slaying of a beautiful theory by an ugly fact. He poetically overstates the case because the rejection of a hypothesis is not a defeat since it raises the need for more appropriate theories and hypotheses and so research continues. If on the other hand the hypothesis is accepted — we did find a relationship between occupation and residential location — this is not the end of the affair, merely a beginning. We need to ask why the hypothesis is correct. Why is there a relationship between occupation and residential location? In this way further avenues of research are opened up, new hypotheses are required and appropriate data and techniques are called for. For a more general hypothesis such as, 'Is the structure of land uses within a city related to distance from the city centre?', a similar process occurs. If we find a particular set of land uses in a certain configuration we need to ask further questions. Is the same pattern found in other cities? Is the pattern consistent through time? What causes this particular configuration? The testing of hypotheses open up the way for more hypotheses, more questions, and it is by answering these questions that our knowledge of the world is increased and improved.

1.3 DATA

In the forthcoming chapters the reader will be introduced to a number of urban data sources. As we have already outlined, it is important that the data selected for analysis are appropriate to the hypotheses which have been raised and the statistical techniques being used to verify those hypotheses. The data which are selected should allow the statistical techniques to answer the questions implied in the hypotheses.

1.3.1 Official data sources

Because of the constraints of time, money and resources, there is a great reliance in urban research on official statistics, e.g. Census, electoral register, rateable value records etc.

The inadequacies of the different official data sources will be mentioned as they are introduced in the course of this book, but at this point it is important to emphasise that official data, like the pronouncements of politicians, should be viewed in a critical manner. Official data sources are often assumed to be completely reliable but students must learn to be aware that, behind the apparent solidity and correctness of official statistics, there often lie considerable sources of error. Before students place their faith in official statistics they must be aware of how the data were collected, since this often highlights the inadequacies of the data. For example the electoral register, which is available in public libraries, lists most persons eligible to vote in a particular town or city. This information is obtained by households returning pre-paid envelopes to the local authority. However, in many cases only one form is delivered per dwelling and, in areas where there are a large number of households in a dwelling (as in old houses which have been subdivided into flats and bedsitters), many households are not recorded. The electoral register thus underestimates the number of households living in multi-occupied dwellings.

Official statistics are collected for specific purposes and to use this data for other than the original purpose requires extreme caution. For example the 1971 Census provides information on the number of households who have moved into areas of the city within the last 5 years preceding the Census. Some researchers have used this variable, expressed as a percentage of the total population in the area, to measure the amount of population ***turnover*** in areas of the city. But this variable does not measure population turnover in areas of expanding population; it measures population ***growth***. How this error arises and the problems it raises for interpretation are discussed in Chapter 5.

Attention has to be paid to data reliability because, even when used with the most subtle hypotheses and the most sophisticated analysis, unreliable data undermine the validity of the whole exercise.

1.4 TECHNIQUES

If data are the raw material of research then techniques are the tools of the trade. They range from the sharp cutting power of regression analysis to the direct simplicity of data assembly and associated mapping. The technique which is used in any exercise should suit the data (e.g. a data set with only 10 observations is not susceptible to complex analysis), and allow the hypotheses to be verified. Students faced with a range of equally suitable techniques should base their selection on the principles of least effort and ease of comprehension and choose the simplest technique. A great deal can be gained from simply mapping data and while the growing sophistication of data analysis often provides extra insights it can become an end in itself. The danger is that we may become so involved in sharpening our tools that we have no time to use them.

1.4.1 Scale and the ecological fallacy

It is important to ensure that the scale of analysis, and the data, are appropriate to the hypotheses under consideration. Many of the hypotheses which we seek to test in urban geography are concerned with the relationships between characteristics of ***individuals***, either persons or households. Unfortunately most of the official data sources, especially the Census, provide information not on individuals, but for ***groups*** of individuals in specific areas of the city. For example if we wish to test the hypothesis that persons in private renting are younger than those in owner-occupation (think of the reasons why this should or should not be true), we need data on the mode of tenure of individuals of different ages. However, from the Census we can only obtain information on the proportion of persons in different tenure categories and the proportion of persons of different ages. If for 10 areas in a city we have a pattern as depicted in ***Figure 1.3***, then we can only ***suggest*** that there is a relationship between the age and tenure of individuals. The data does not prove it, since it may be that the person aged 15–29 years may not necessarily be in private renting. To make conclusions at the individual scale from analysis of data at the aggregate scale, in this case to make conclusions at the individual level from the analysis of data referring to population in ten areas, is to make the 'ecological fallacy'.

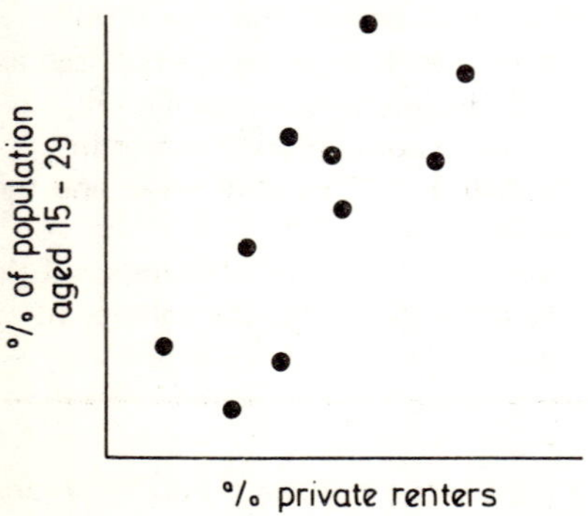

Figure 1.3 Hypothetical relationship between tenure and age

The dangers of the ecological fallacy were dramatically shown by Robinson (1950). In a study of illiteracy in the United States he found that at the level of the Census Bureau's nine geographic divisions of the United States, there was a strong relationship between the per cent illiterate and the per cent negro. The ecological correlation was +0.946 (1.000 = complete positive correlation). When the same variables were used for states, the correlation dropped to +0.773. But when the colour and illiteracy characteristics of individuals were computed, the correlation was found to be only +0.203. If he had only used the ecological data, for the Census division or the states, and assumed that results from ecological analysis can be used to make conclusions about individuals, then Robinson would have believed that there was a strong relationship between a person's colour and their reading and writing ability. However, as the correlation for individual data shows, there is only a very weak relationship between a person's colour and his reading/writing ability.

The problem of scale still arises even when we are not dealing with individual characteristics. For example if we are testing the relationship between house density and rateable value, it is important to remember that our results, and therefore our conclusions, are scale specific. For example if we use data on density and rateable value which refers to 1-km squares – i.e. the number of houses in this area and the average rateable value of all houses in this area – this does not necessarily mean that the same results will be found for 2-km, 5-km or 10-km squares. Before we can make general conclusions it is important that the relationship between any two variables is tested at a variety of scales.

Most of the data which will be presented in this book refer to area units. The maintenance of confidentiality restricts the release of data on individuals from official sources. The ecological fallacy means that we cannot use this data to assume that the relationships between variables for spatial units also applies at the individual level. We can only suggest that the relationship remains constant. This applies to other data referring to specific spatial scales. We cannot assume that the relationships we discover at one scale applies to other scales.

1.4.2 Techniques and hypotheses

In terms of the relationship between statistical techniques and hypothesis verification, an important distinction has to be made between causal and statistical relationships. In a causal hypothesis we suggest that variable A affects (causes) variable B or vice versa. However, we cannot verify this relationship by only noting the statistical relationship, because a statistical relationship or association can be found between two variables even when there is no causal link. For example a most interesting statistical relationship exists between the number of prostitutes arrested yearly in Australian cities and the number of divinity students graduating from British universities and colleges. The correlation coefficient for the post-war period is +0.91, a statistically significant result. This example shows that a statistical relationship, even when tested for significance, can be meaningless in causal terms. Alternatively, the strong statistical association we find between any two variables may be due to causal effects of a third factor. For example there is an almost perfect correlation (+0.99) between the number of marriages and the sale of ball-point pens in the United States for the period 1945–1950. Both of these variables are due to the changed circumstances after the war, on the one hand the return of service personnel and a return to normality in human relations and on the other the increase in consumer demand, but they are not themselves causally related.

The use of statistical techniques does not provide a perfect method of verifying hypotheses. This means that the acceptance or rejection of a hypothesis relies in large measure on individual judgement and belief. We accept the results of some statistical tests, but not others, e.g. the prostitutes/divinity students example, because we ***believe*** them to be right. Behind the apparent objectivity of statistical analysis lies a great deal of subjective judgement, and it is important to remember that statistical techniques, like data sources, are an aid to critical thought, not a replacement.

CHAPTER 2 THE URBAN SYSTEM

2.1 INTRODUCTION

Towns and cities do not exist in isolation. Together they form an urban system. In this chapter we will focus on the range of data and techniques which permit an analysis of urban areas as parts of a national or regional urban system. Two themes will be developed; the character of the urban hierarchy and the differences between urban areas in relation to a number of characteristics.

2.2 THE URBAN HIERARCHY

By the term ***urban hierarchy*** we simply mean that within one country or one region there are a few large cities and a greater number of smaller cities and towns. The nature of the urban hierarchy can be depicted by using double logarithmic graph paper and plotting the population size of settlements on the vertical scale and the rank order on the horizontal scale. Rank order refers to the position of a city with respect to the population sizes of the other cities. The largest city has a rank of 1, the next largest has a rank of 2. . . . . . . and so on.

Figure 2.1 shows three types of relationship between population size and rank using this graphical method. In ***Figure 2.1a*** there is a straight line relationship between rank and size.

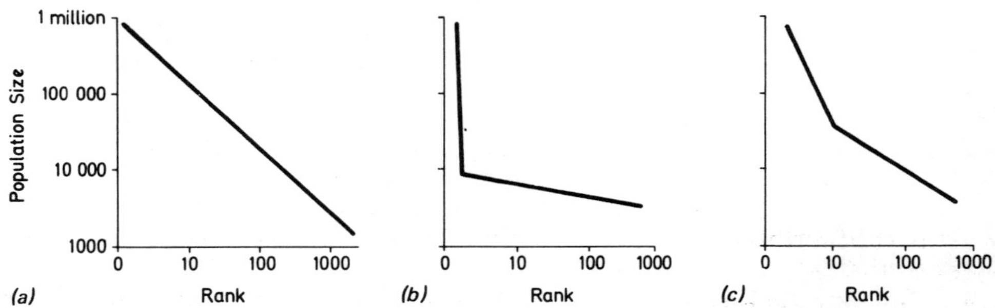

Figure 2.1 Three rank–size distributions

This ***regular distribution***, as it has been termed, occurs when there is a continuous relationship between size and rank. This relationship can be expressed in the ***rank–size rule formula*** where,

$$P_n = \frac{P_1}{n}$$

where P_n is the population of the nth town and P_1 is the population of the largest city. If there is a regular distribution between rank and population size then if the largest city has a population of 100 000(P_1), the fifth ranked town should have a population of,

$$P_5 = \frac{100\,000}{5}$$

$$= 20\,000$$

In the case of ***primate distributions*** (***see Figure 2.1b***) the largest city, which is termed the primate city, dominates the urban hierarchy. In a primate distribution the largest city has a proportionally larger population than in the case of the regular distribution. Between the two extremes of regular and primate distributions there are a variety of ***intermediate distributions. Figure 2.1c*** provides an example of an intermediate distribution wherein the larger cities dominate the hierarchy; they have a proportionally larger population than that of the larger towns in a regular distribution.

The three cases shown in ***Figure 2.1*** reflect differential rates of growth within an urban hierarchy. In the regular distribution population growth, or decline, would seem to be uniform throughout the urban hierarchy while in the primate distribution one city has clearly grown much larger than all the other cities or, alternatively, has declined less than the other urban areas. In comparison to the regular distribution the intermediate distribution shown in ***Figure 2.1c*** suggests that the larger cities have grown much larger or declined less than the medium-sized towns.

2.3 INTERNATIONAL COMPARISONS

2.3.1 Rank–size distribution

A number of studies (e.g. Berry, 1961) have examined the rank–size distribution of urban hierarchies in different countries. Such studies have suggested that, in countries which are larger than average, have a long history of urbanisation and which are also economically and politically complex (e.g. the United States), the rank–size relationship exhibits a regular distribution. But in countries which are smaller than average, have a short history of urbanisation, such as Thailand, then the rank–size relationship shows a primate distribution.

Primate distributions are characteristic of the urban hierarchies of underdeveloped countries. This primacy occurs and is maintained because of the patterns of population and economic growth. Most underdeveloped countries are experiencing large increases in population but, because the better levels of health provision in the largest city reduce mortality levels lower than those found in other parts of the country, a proportionally larger share of this population growth occurs in the primate city. Equally important is the pattern of economic development. The small amounts of economic development which have occurred in the underdeveloped countries have very largely taken place in the largest cities. There is a definite urban bias to the pattern of economic growth and such growth provides a powerful pull effect in rural–urban migration, which further increase the population size of the primate city. The urban bias of economic growth in underdeveloped countries becomes a self-fulfilling process; economic growth takes place in the primate city and these job opportunities attract migrants and this population growth in turn creates job opportunities in the service sector and so more migrants are attracted to the big city.

2.3.2 The rank–size rule

Although it is useful to plot the rank–size relationship for a sample of countries you may find it difficult, if not impossible, to obtain the relevant data. Even if you were able to obtain the relevant data, the number of observations makes the plotting of rank and population size a long and time-consuming business. However, a useful reference source is the ***United Nations Demographic Yearbook***. This yearbook, which is published by the Department of Economic and Social Affairs of the United Nations, provides a summary of urban populations in the member countries of the United Nations. Since 1948 it has provided a table which lists the population of cities over 100 000 inhabitants by country for most countries of the world. This data source can be used to compare the differences between countries in the rate of urban population growth and to note the differences between different types of countries in terms of the percentage of the national populations living in very large cities. Apart from providing an invaluable reference source this data source can also be used to compare the nature of the upper levels of the urban hierarchies in different countries of the world.

According to the rank–size rule the second largest city in a country should have approximately half the population of the primate city. If we return to our earlier example where the primate city had a population of 100 000 then, if the urban hierarchy exhibits a regular distribution, the population of the second largest city should be

$$P_2 = \frac{P_1}{2}$$

$$= \frac{100\,000}{2}$$

$$= 50\,000$$

In this case the second largest city has 0.5 of the population of the primate city. If the urban hierarchy has a regular distribution then the population of the second largest city expressed as a fraction of the primate city should be 0.5. The closer this fraction is to 0.5 the more regular is the urban hierarchy and conversely the smaller this fraction the more the urban hierarchy displays a primate distribution.

We can use the data in the previously noted table of the ***United Nations Demographic Yearbook*** to calculate the population of the second largest city as a fraction of the primate city for a variety of different countries. ***Table 2.1*** lists this fraction for the United States, Argentina and Thailand taken from the data in the 1962 and 1976 ***Demographic Yearbooks***. Note that the figures for the United States are very close to 0.5 indicating a regular distribution in the urban hierarchy. The figures for Thailand and Argentina indicate a primate distribution. Do the figures for 1962 and 1976 for these latter two countries indicate that primacy is increasing or decreasing? Calculate this fraction for a sample of other countries, compare them with the values in ***Table 2.1*** and discuss them in the light of the remarks made on the rank–size distributions in different types of countries.

TABLE 2.1 SECOND LARGEST CITY AS FRACTION OF PRIMATE CITY: INTERNATIONAL COMPARISONS

	United States	Argentina	Thailand
1962	0.45	0.09	0.28
1976	0.41	0.26	0.33

2.4 THE URBAN HIERARCHY IN BRITAIN

2.4.1 Data sources

Apart from 1941 there has been a Census of the population in Britain every 10 years since 1801. The successive Censuses published separately for England and Wales and for Scotland list the population of each city, town and rural area. The Censuses of 1801, 1811, 1821 and 1831 suffer from large errors, since the population count was spread over a number of days which increased the possibility of double counting. The data from the 1841 Census onwards are considered to be much more reliable as the enumerators – the people who actually counted the population – were required to undertake the population count in 1 day. Despite the errors of the earlier Censuses the successive ones provide a useful data source for the examination of the changing nature of the urban hierarchy in Britain.

2.4.2 Analysis of historical patterns

The Census data can be used to plot the rank–size relationship for all urban areas in Britain from 1801 to the present day. In ***Figure 2.2***, for example, the rank–size relationship has been plotted for all towns and cities in England and Wales with a population of over 2500 for the periods 1801, 1851 and 1911. The relationship portrayed in ***Figure 2.2*** is similar to the intermediate distribution shown in ***Figure 2.1***. In England and Wales during this period the larger towns and cities dominated the urban system and ***Figure 2.2*** suggests that a number of larger cities have been 'grafted' on to a regular distribution. This pattern could be explained with reference to the enormous growth of large cities, especially London, which grew up as international centres of commerce and industry within a smaller than average country. It is also interesting to note that while the successive Censuses have recorded increases in population for all the urban areas the form of the rank–size relationship shown in ***Figure 2.2*** has remained remarkably constant over the 110-year period.

This Census data can also be used to plot the rank–size relationship at both the national and the regional scale. At the national scale the data could be used to extend the period covered by ***Figure 2.2***; for example, what is the form of the rank–size relationship for towns with more than 10 000 inhabitants in England and Wales from 1911 to the present day? Alternatively, the rank–size relationship for all the urban settlements in one particular region could be plotted for a number of time periods. As an exercise plot the rank–size relationship for all towns with over 5000 population in your county or region for a number of periods between 1801 and 1971. What is the form of the relationship? Has it changed over this time?

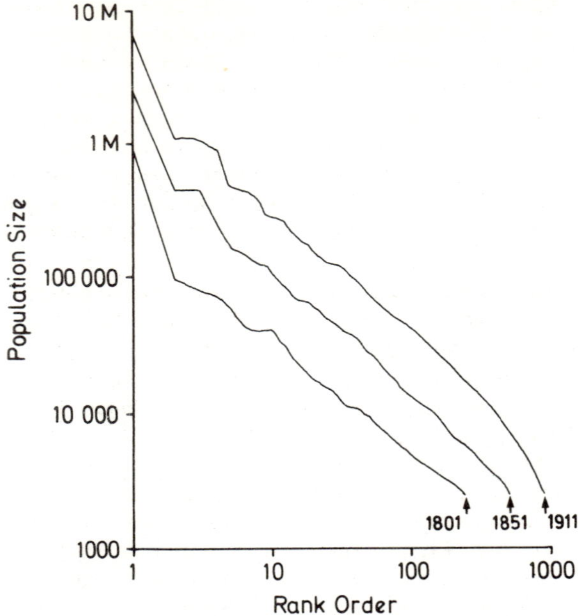

Figure 2.2 Rank–size relationship for towns in England and Wales; 1801, 1851 and 1911 (After Robson, 1973)

2.4.3 Analysis of contemporary patterns

It is often useful to know the changing position of one particular city in the urban hierarchy. This examination can be based on the information in the Censuses of 1951, 1961 and 1971. The County Reports of each of these Censuses list the population of each local authority area in the respective counties. Alternatively, the 1971 Census, Preliminary Reports, published for England and Wales and for Scotland, which give the provisional figures for the population change between 1951 and 1971 for each town with a population greater than 50 000 could be used. Although this latter source contains only provisional estimates, it has the advantage of providing all the necessary information in one single volume. This latter source has been used to determine the rank of 25 cities in England and Wales from 1951 to

TABLE 2.2 RANK OF 25 CITIES IN ENGLAND AND WALES

City	Rank 1971	Rank 1951
London	1	1
Birmingham	2	2
Liverpool	3	3
Manchester	4	4
Sheffield	5	5
Leeds	6	6
Bristol	7	7
Teeside	8	8
Coventry	9	15
Nottingham	10	9
Bradford	11	11
Hull	12	10
Leicester	13	13
Cardiff	14	16
Wolverhampton	15	17
Stoke	16	14
Plymouth	17	19
Newcastle	18	12
Derby	19	21
Sunderland	20	20
Southampton	21	22
Portsmouth	22	18
Dudley	23	25
Walsall	24	23
Swansea	25	24

1971 as shown in ***Table 2.2***. This table shows that, while the top eight cities have retained their relative position in the 20-year period, there have been significant changes in the relative rank of some medium-sized cities. In ***Figure 2.3*** the changing rank position of Coventry and Newcastle are depicted. In 1951 Newcastle, with a population of 291 724, was ranked 12 but by 1971 the population had fallen to 222 209 and its rank had dropped to 18. In 1951 Coventry had a population of 265 765 and a rank of 15 but by 1971 the population had risen to 335 228 and its rank position had climbed to 9. These respective changes are a function of the changing economic fortunes of the two cities. Newcastle, with its reliance on the more traditional industries, such as shipbuilding, has witnessed a steady decline in jobs and associated high levels of out-migration. Coventry, by contrast, has been one of the main centres for the expansion of the car assembly industry which has created job opportunities and attracted many migrants from other parts of the country.

The data shown in ***Table 2.2*** can be examined in two other ways. First, the changing economic fortunes of cities in different parts of the country are reflected in the changes in rank. To highlight this feature plot the distribution of cities which have respectively increased and decreased in rank position. Is there any noticeable spatial pattern? In other words, are the cities which have increased, or decreased, in rank position located in

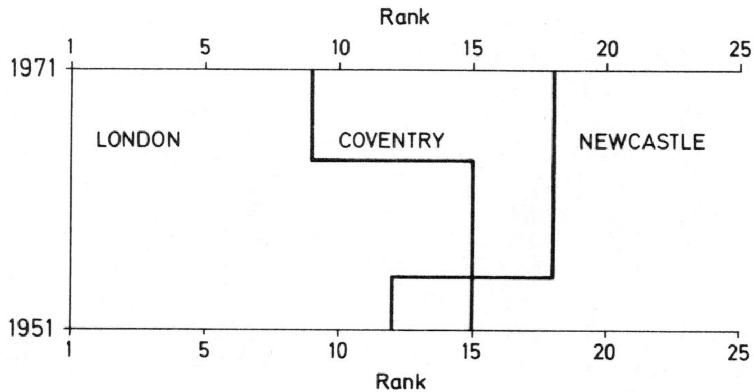

Figure 2.3 Change in rank position of Newcastle and Coventry

TABLE 2.3 POPULATION REDISTRIBUTION IN LANCASHIRE AND SUSSEX

Administrative units	Lancashire		Sussex	
	1951	1971	1951	1971
County Boroughs	60.1	51.0	45.9	41.2
Urban and Municipal Districts	34.1	40.3	29.1	30.9
Rural Districts	5.8	8.7	25.0	27.9
TOTAL	100.0	100.0	100.0	100.0

particular parts of the country? Discuss this spatial pattern with reference to your knowledge of the pattern of economic growth in post-war England and Wales.

Secondly, to test the relationship between rank positions in 1951 and in 1971 a ***Spearman Rank Correlation Coefficient*** (R_s) could be calculated. (A much fuller explanation of this statistic is given in the next chapter.) The R_s value is obtained in the following formula:

$$R_s \quad 1 - \frac{6 \times \Sigma D^2}{N^3 - N}$$

where D^2 is the squared difference between the rank of a city in 1951 and the rank in 1971 and N refers to the number of pairs. In this case N is equal to 25. For each city calculate D^2 by subtracting the 1971 rank from the 1951 rank and squaring the result. For example, Hull which was ranked 12 in 1971 and 10 in 1951 has a D^2 value of 4. Add up all the D^2 values and calculate R_s. With 25 pairs, significant values of R_s must equal or exceed ±0.336 at the 95% confidence level. Is your computed R_s value statistically significant? If so what conclusions can you draw concerning the stability of rank positions between 1951 and 1971?

Since the Census Preliminary Reports contain information on towns with a population greater than 50 000 the preceding forms of analysis could be extended to include more than the top 25 cities. Alternatively, the analysis could be pitched at the regional scale. Note the changing rank of all the towns over 50 000 in your county or region. Have there been any significant changes? Is there any spatial variation to these changes?

2.4.4 Population redistribution

The previously quoted Census sources can be used to note the changing distribution of the population within urban hierarchies. The Preliminary Reports of the Census provide information on the distribution of county populations by county boroughs, urban and municipal boroughs, and rural districts. The county boroughs are the very large towns and cities and the urban and municipal boroughs are the smaller towns with a population greater than 15 000. From the data in the Preliminary Reports, consider the case of two counties, Lancashire and Sussex. The percentage of the respective county populations in each of these administrative units in 1951 and 1971 are shown in ***Table 2.3***. In both counties there has been a decline in the percentage of the population in the county boroughs and an increase in the percentage of the population living in the smaller towns and rural areas. In effect there has been a redistribution of the population in both counties towards the lower end of the urban hierarchy. This redistribution of the population is a function of the growing surburbanisation of the population as households move out to live beyond the boundaries of the major towns and cities.

To note the spatial patterns of population redistribution it is necessary to use the more detailed information in the County Reports of the Census. For example, you may wish to

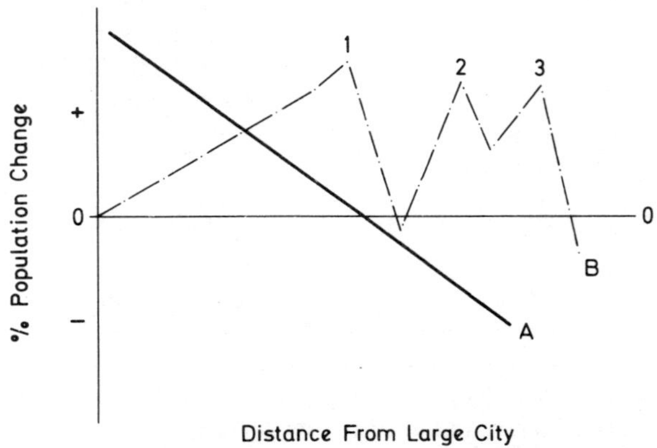

Figure 2.4 Population change with distance from large city

examine the spatial distribution of population change in areas which cross a transect drawn from the largest city in a particular city region. Such an exercise would be based on the data provided by the County Reports of the 1951, 1961 and 1971 Census. These sources would enable you to calculate the level of population change in particular administrative units.

Figure 2.4 provides an example of two alternative patterns of population change in areas around a large city. In the case of A in the figure, population growth is greatest in those areas closest to the city and declines markedly beyond a certain distance. This distance could be the range of commuting to the city. However, the pattern of population change may not be a simple linear function of distance. In the case of B there is no clear-cut relationship between population change and distance from a large city. This second pattern may arise in a number of circumstances. Planning regulations may restrict the construction of new dwellings to certain areas — in this case areas marked 1, 2 and 3 on the diagram. Planning regulations may also have directed industrial development to certain favoured sites. If the planning authorities had created a ***green belt*** around a city draw the general form of the relationship you would expect to find between population growth and distance from the large city. As an exercise draw a transect from a large city in your local region up to a distance of 50 km. From the relevant County Report(s) of the 1951 and 1971 Censuses calculate the population change for all urban and municipal boroughs and rural districts which lie on this transect over the period 1951 and 1971. Plot the distribution of population change against distance as shown in ***Figure 2.4***. Does the pattern correspond with either the A or B pattern shown in ***Figure 2.4***? Draw transects in different directions and perform the same kind of analysis. Are there any differences between the patterns recorded for the different transects?

Appendix A contains data and maps which allow an examination of population change and redistribution in two very different types of urban regions.

2.4.5 The size and spacing of settlements

The urban hierarchy is composed of a small number of large cities and a greater number of smaller cities and towns. The empirical regularity between the size and rank of settlements is matched by the relationship between the size and spacing of settlements. Put simply, cities and larger towns are spaced further apart than smaller towns and villages. This pattern has been noted in a number of studies. For example Christaller (***see*** Berry, 1967) identified

Figure 2.5 The size and spacing of medium-sized towns in England and Wales (After House, 1955)

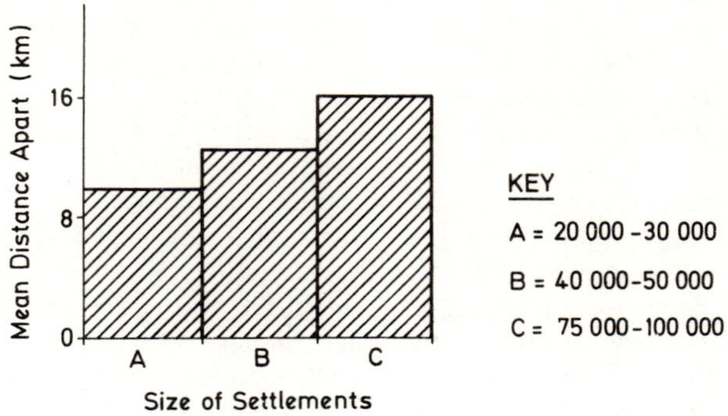

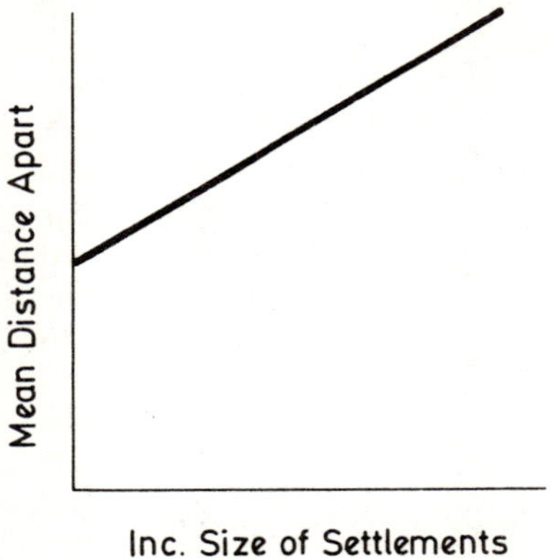

Figure 2.6 The general relationship between spacing and size of settlements

four levels to the urban hierarchy in southern Germany and, in ascending order, he found that ***villages*** were located 7 km apart, ***county towns*** were 21 km apart, ***provincial cities*** were 108 km apart and the larger ***regional cities*** were 178 km apart. Brush and Bracey (1955) identified a two-level hierarchy in southern England with larger towns located 33.8 km apart and smaller towns spaced between 13 and 16 km apart. Finally, House (1953) examined the spacing between medium-sized cities in England and Wales. His results are portrayed in ***Figure 2.5***. We can generalise from these results to suggest, as shown in ***Figure 2.6***, that distance between towns of a certain size will increase as the population size of towns increase. Calibrate the axes of ***Figure 2.6*** for a variety of different size categories, using the data provided by the Census and relevant maps, with reference to both a national and a regional analysis of urban areas.

2.4.6 Spheres of influence

Towns and cities differ with respect to the range of services and facilities which they offer to consumers. Larger towns and cities contain the range of everyday goods and services, such as bread shops, which are also found in smaller towns and villages, in addition to a range of special and luxury goods and services, such as a jeweller's shop. These latter goods and services, which are termed ***higher order goods and services***, are only found in the larger towns and cities since it is only these areas which have a population large enough to support such specialised shops and services. For example if we assume that, out of every 1000 people, only 3 are likely to go to a jeweller's shop in any one month, then in a small town of 10 000 people a jeweller can only expect to get, on average, 30 customers per month. In contrast, a jeweller in a large town with a population of 60 000 can expect to get 180 customers each month. It is, therefore, more economically viable for a jeweller to set up his/her shop in the larger town; this principle applies to the range of higher order goods and services. You can test this principle by selecting a range of higher order goods and services and noting their location within the urban hierarchy. A useful higher order service to examine is the large department store e.g. Woolworths, Marks and Spencers, etc. For a range of different sized towns and cities in your local region plot on a graph the relationship between the number of department stores in each settlement and the population size of the settlement. Are there any significant 'break-points' which indicate that, below a certain population size, department stores will not be found? Such 'break-points' are termed the ***threshold*** of a particular good or service (***see*** Chapter 6). Identify the threshold value for a range of goods and services and comment on your results.

The fact that larger towns and cities have a range of specialised goods and services not found in smaller settlements suggests that these larger settlements will attract consumers from areas beyond the city boundary. These areas have been variously referred to as ***catchment areas***, ***hinterlands*** and ***spheres of influence. The larger the town or city, the greater the number of specialised goods and services and hence the larger the sphere of influence.***

The extent of a particular town's sphere of influence can be identified by a number of measures. Carter (1972), for example, has used a variety of criteria to delimit the sphere of influence of Aberystwyth as shown in ***Figure 2.7***. Although there are some slight differences in the areas demarcated by the different criteria, there is a broad level of agreement over the extent of Aberystwyth's sphere of influence.

The sphere of influence of a town or city can also be delimited by questionnaire studies which ask a sample of households in small towns and rural areas which town they go to for specialised goods such as furniture, and/or specialised services such as a solicitor. The results of such a survey enable spheres of influence to be identified and mapped. Questionnaire surveys provide the best source of information, but by their very nature

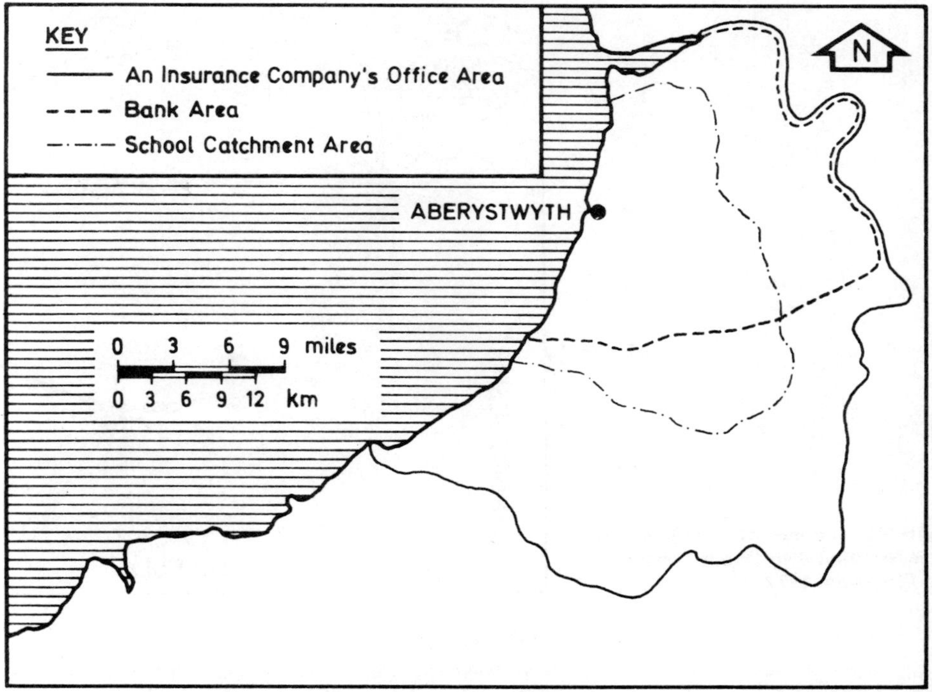

Figure 2.7 Aberystwyth's spheres of influence (From Carter, 1972)

they are expensive and time-consuming exercises. An alternative method has been used recently by Everson and Fitzgerald (1972). They have plotted the distribution of places regularly mentioned in the news section and advertising sections of three local newspapers in East Anglia, as shown in ***Figure 2.8***, and suggest that local newspapers used in this way provide a quick method for identifying spheres of influence. This method and source, though by no means perfect, does provide a cheaper and quicker alternative to questionnaire surveys.

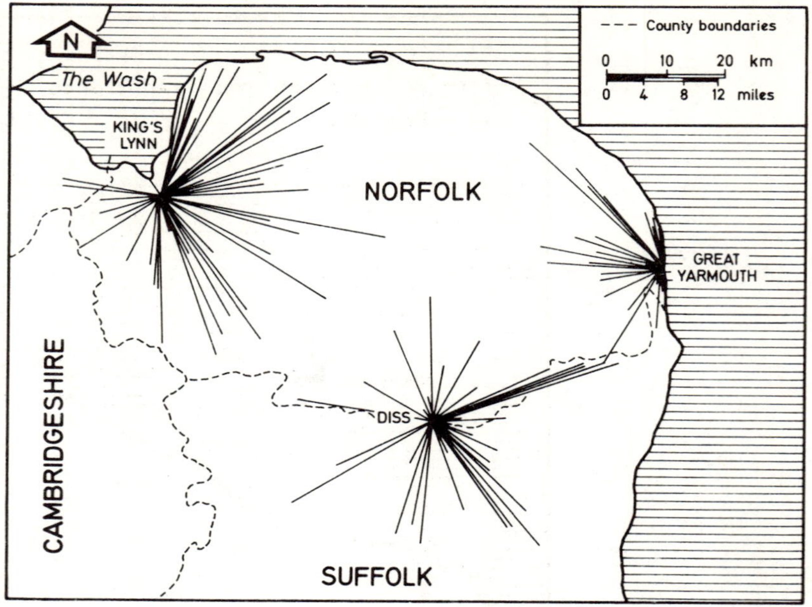

Figure 2.8 The spheres of influence of three Norfolk towns (From Everson and Fitzgerald, 1972)

Reilly's Law

Whatever the method used to delimit spheres of influence the actual results can be compared with the results predicted by Reilly's Law. Reilly argued that a town's sphere of influence with regard to shopping patterns is a function of the population of a town and its distance to other towns. More specifically, Reilly argued that the boundary between the spheres of influence of two towns could be predicted by the following formula:

$$B_j = \frac{D_{ij}}{1 + \sqrt{P_i/P_j}}$$

where, B_j is the predicted distance from town j to the boundary between its sphere of influence and that of town i, D_{ij} is the distance from town i to town j, P_i is the population of town i and P_j is the population of town j. This equation when applied to the situation in *Figure 2.9* suggests that shoppers will travel up to 5.85 km to town j, but people living beyond this point will travel to shop in town i. Calculate the distance of an actual boundary between two towns' spheres of influence and compare it with the predicted distance according to Reilly's Law. Is there any variation?

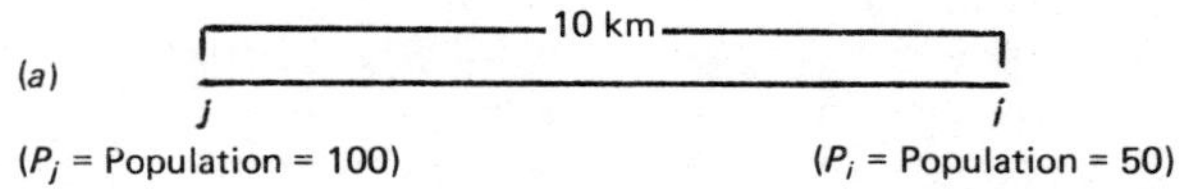

$$B_j = \frac{D_{ij}}{1 + \sqrt{P_i/P_j}}$$

$$= \frac{10}{1 + \sqrt{50/100}}$$

$$= 5.85$$

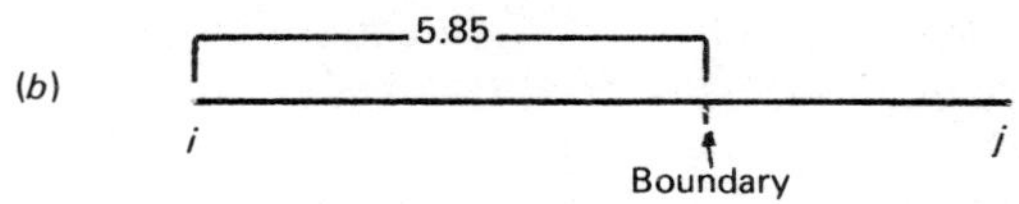

Figure 2.9 Reilly's Law and spheres of influence

2.5 VARIATIONS WITHIN THE URBAN SYSTEM

2.5.1 Introduction

Towns and cities differ in terms of their size, rank, spacing and the extent of their spheres of influence. They also differ with respect to a number of other characteristics. In this section we will consider some data sources which allow a comparison between towns and cities in Britain across a number of attributes.

2.5.2 Population change

As we have already noted, the 1971 Census, Preliminary Reports provide a useful source of information relating to population change for all major cities in Britain between 1951 and 1971. The England and Wales Preliminary Report for 1971 has been used by Kivell (1972) to compare the variation in the growth rates of metropolitan areas over the period 1961 to 1971. These metropolitan areas, identified by Kivell on the basis of the number of persons employed in non-agricultural employment, were divided into inner and outer areas. The inner metropolitan areas were defined as the core areas of the respective city regions and the outer metropolitan areas were defined as those areas where a majority of the population commuted to the core area for employment. The variation between the population changes in the inner and outer areas of each of the metropolitan areas is shown in ***Figure 2.10***. The most obvious conclusion to draw from this diagram is that there have been differential rates of growth between the inner and outer areas of metropolitan areas with the greatest population increases being recorded for the outer metropolitan areas. In almost all cases the outer areas recorded greater population growth than the inner areas and in a majority of cases the inner areas witnessed population decline. The results suggest that there has been a substantial decentralisation of the population over the 10-year period. The results also suggest that this decentralisation has not been uniform throughout the country. The greatest gains for outer metropolitan areas have been recorded for the expanding cities of south and south-east England, while the greatest losses for inner metropolitan areas have occurred in the cities of the north and north-west.

2.5.3 Housing

Information on the number of dwellings in the different tenure categories (owner-occupation private renting and local authority) and the number of dwellings defined as unfit or lacking amenities for each local authority is available in the respective County Reports of the 1961 and 1971 Census. Since this information is contained in a large number of separate volumes comparative analysis can be very time-consuming. However, there are two sources which allow a quick comparison of the housing characteristics of different towns and cities in England and Wales. The ***Local Housing Statistics*** are published every 3 months and for each local authority they give the number of dwellings under construction and completed in the three different tenure categories. This source also gives information on the number of dwellings demolished and the number of housing-improvement grants approved in each local

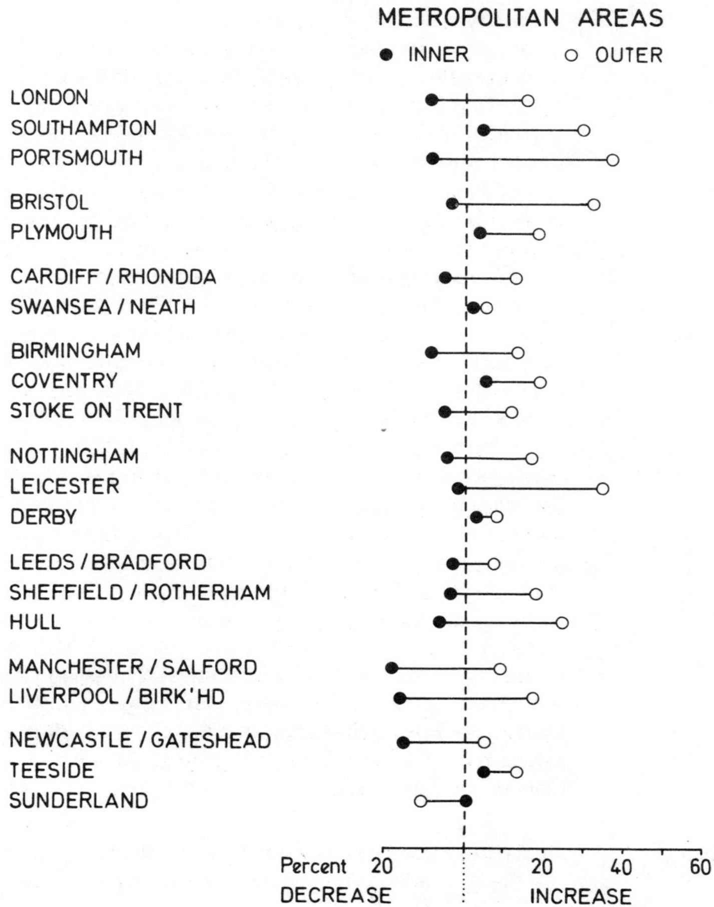

Figure 2.10 Population change in metropolitan areas, 1961–1971 (From Kivell, 1972)

authority area. Since this source also gives the estimated population of the local authority area the statistics are generally expressed per 1000 household or per 1000 persons. Such statistics allow us to examine the often large variation between local authorities in relation to housing activity. For example between 1966 and 1972 building commenced on 121 local authority houses per 1000 households in Manchester while the comparable figure for Cardiff was only 20 per 1000. In the same period 152 dwellings per 1000 households were demolished in Manchester but only three per 1000 in Plymouth. An examination of the data in the ***Local Housing Statistics*** enables you to note which local authorities demolished most houses, built most council houses and approved most grants in any time period.

The ***Housing Statistics***, which are published annually by the Chartered Institute of Public Finance and Accountancy (before 1972 this body was known as the Institute of Municipal Treasurers and Accountants), provide information on the age of housing, capital expenditure on local authority housing and the rents of local authority dwellings for the majority of local authorities in England and Wales. This information is obtained by the response of local authorities to a questionnaire administered by the aforementioned Institute. Since there is no statutory obligation to provide this information, not all local authorities comply. However, all the large local authorities do provide the requested information. The ***Housing Statistics*** are published in two parts. Part 1 contains information on the average weekly rents of different types of local authority housing and the number of rent rebates and allowances administered by each local authority. Part 2 provides a breakdown of the capital expenditure and income relating to local authority housing in each local authority. These statistics can be used to examine the variation in local authority housing rents between different local authorities. For example while a local authority tenant of a two-bedroomed house built between 1945 and 1965 was charged £2.95 weekly rent in Rochdale in 1974, the tenant of a similar property was charged £4.65 in Brighton and £5.04 in Southend. Using the ***Housing Statistics*** identify the local authority which charges the highest and lowest rents for different types of local authority housing. Examine the range between the highest and lowest rents.

2.5.4 Population and social services

Apart from the 1971 Census; Persons of Pensionable Age Table there is no single volume of the Census which provides information on the population characteristics of different towns and cities in Britain. However, the ***Personal Social Services Statistics***, another publication of the Institute of Public Finance and Accountancy, does provide such a source. For most local authorities in England and Wales this source provides figures for the percentage of the population under 5 years of age, the percentage of the population under 18 years of age, the percentage of the population 65 years of age and over and the percentage of the population 75 years of age and over. For example in 1971 9.4% of the population in Coventry were 65 years of age or more while the corresponding figure for Bournemouth was 28.8%. Can you explain the difference in demographic structure between the two cities?

The ***Personal Social Services Statistics*** also provide information concerning the expenditure by local authorities on different types of welfare provision. Such information includes the total expenditure on social services by each local authority and a breakdown of this expenditure into different categories such as the amount spent on child clinics, homes for elderly people etc. For example in 1976 Manchester spent £35 614 per 1000 persons on social services, Islington spent £61 159 and Dudley only spent £12 759. In ***Table 2.4*** three items have been taken from the ***Personal Social Services Statistics*** for 25 cities in England and Wales. Calculate the mean of each item and note the variations from the mean. How would you measure the degree of association between the three items?

It is important to remember that when we are dealing with welfare expenditure, such as the amount spent on social services and local authority housing, we are in fact dealing with the outcome of political decisions. Welfare expenditure is ultimately governed by the central authorities but it is administered by the local authorities. A number of studies have suggested that variations in welfare provision between local authorities is primarily a reflection of differences in the local political structure, with local authorities where the Labour party is in power spending more on welfare than local authorities where the Conservatives are in power (Boaden, 1971).

If your local authority has experienced a change in the political party in power note the preceding and subsequent levels of welfare expenditure. Compare the welfare expenditure of local authorities in your region with different political parties in power. Are there any differences?

In an important book Smith (1977) has argued that the main focus for human geography should be the analysis of welfare: ***who gets what, where and how***. Although you may not be

TABLE 2.4 AGE CHARACTERISTICS AND WELFARE EXPENDITURE OF 25 CITIES

City	A	B	C
London	12.9	5.1	10.1
Birmingham	12.7	2.3	0.0
Liverpool	14.0	1.1	26.9
Manchester	14.0	3.0	16.0
Sheffield	13.9	3.6	35.8
Leeds	13.6	2.9	5.8
Bristol	14.1	10.6	28.4
Teeside	9.9	2.7	0.0
Coventry	9.4	3.0	4.4
Nottingham	12.5	2.1	0.0
Bradford	13.6	6.7	0.0
Hull	12.2	2.2	47.5
Leicester	13.7	1.1	0.0
Cardiff	13.3	2.5	37.7
Wolverhampton	10.3	2.4	0.0
Stoke	11.3	3.7	6.5
Plymouth	14.7	1.2	0.0
Newcastle	15.7	3.4	0.0
Derby	13.7	1.1	0.0
Sunderland	11.9	2.4	3.2
Southampton	11.9	3.1	13.6
Portsmouth	18.0	3.8	0.0
Dudley	10.2	3.4	0.0
Walsall	10.7	2.2	0.0
Swansea	12.9	3.4	0.0

A Percentage of population 65 years of age and over
B Local authority expenditure (£'s) on child clinics and centres per child under 5 years of age
C Local authority expenditure (£'s) on family planning per 1000 people

able to answer all of these questions an examination of the ***Local Housing Statistics***, the ***Housing Statistics*** and the ***Personal Social Services Statistics*** do provide a basis for discussing some of these fundamental questions.

CHAPTER 3 THE CENSUS AND THE INTERNAL STRUCTURE OF THE CITY

3.1 SMALL AREA STATISTICS

The most comprehensive source of contemporary information on the internal structure of British cities is the 1971 Census of Population, statistics for small areas. Comparable information is also available for 1961 and 1966. This census information is provided at two spatial scales. The smallest scale is the enumeration district level. Enumeration districts are the areas assigned to enumerators who were employed to distribute census forms to each household, collect these forms and check any answers which are missing, incomplete or incorrectly answered. Each enumeration district includes, on average, 150 households or 500 persons. For the 1971 Census there were approximately 120 000 enumeration districts in England, Wales and Scotland. The same information, as provided at the enumeration district level is also available for wards of the city. Wards are political units, each ward is represented by two to three councillors on the city council and can vary in population size from 8000 to 30 000. The smallest wards are inner city wards which have experienced heavy population loss and the largest are the peripheral wards which have experienced population growth in the last 10–20 years.

The differences in the spatial coverage of these two area units are shown with respect to Bristol in ***Figure 3.1***. In Bristol, which had a total population of 426 657 in 1971 there were 821 enumeration districts for the 1971 Census and 28 wards. In terms of detecting small area differences and reducing internal variations, the enumeration districts are the best units of observation. But, the very large number of enumeration districts implies that only those with either access to computer facilities or a cohort of research assistants can adequately deal with all the enumeration districts in a large or even medium-sized city. For those who are interested in very small areas of the city enumeration districts do, of course, provide the most appropriate data base, but if we are interested in the internal structure of the whole city and have limited computer facilities then the wards provide the most convenient observation unit. An added advantage in using wards is that, unlike enumeration districts, the boundaries have remained consistent in the 1961, 1966 and the 1971 Censuses. Comparisons can, therefore, be made across a 10- and 5-year period.

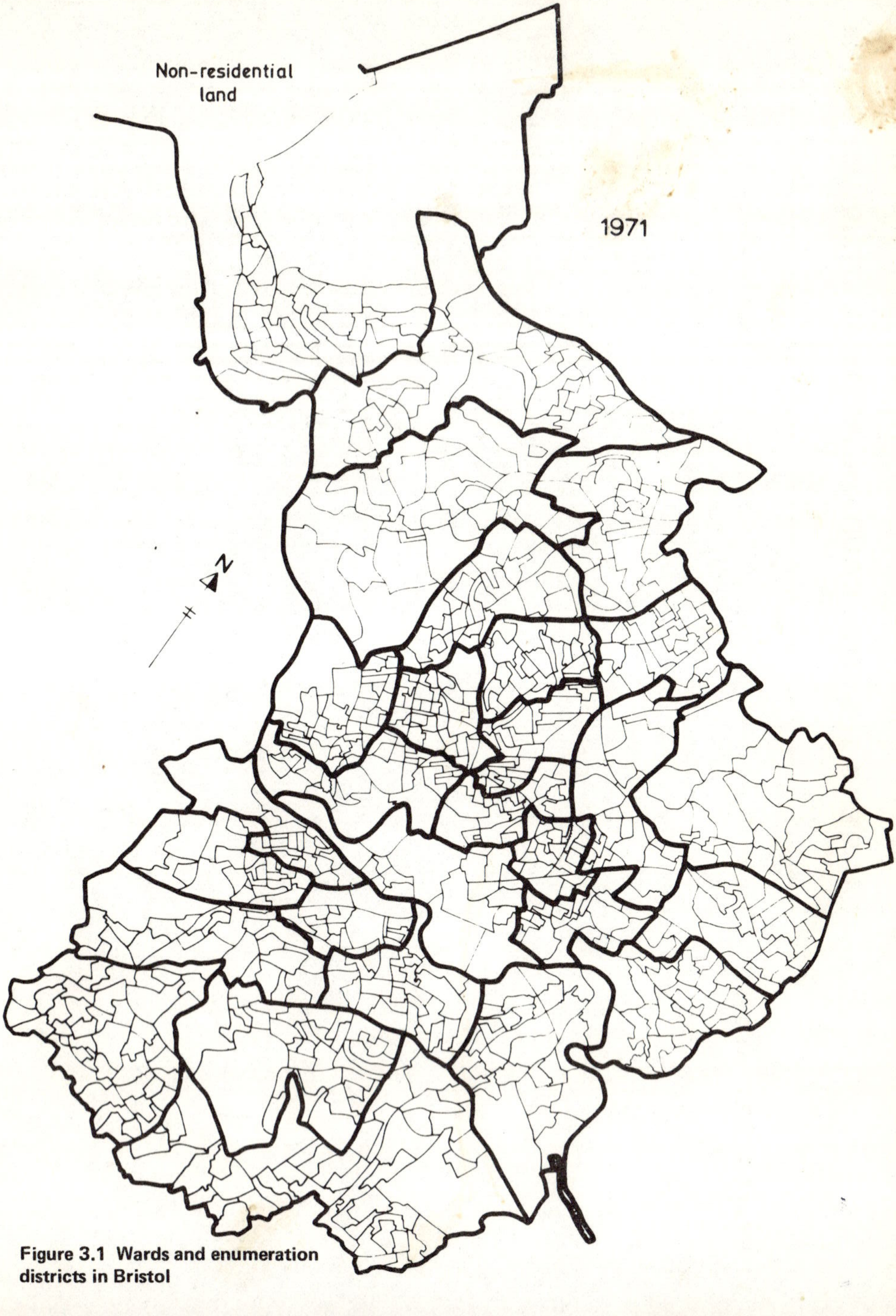

Figure 3.1 Wards and enumeration districts in Bristol

3.1.1 Data base

Having identified the most suitable data unit we now have to consider the type and range of data available at this spatial scale. The type of information which is available is shown in ***Tables 3.1*, *3.2*** and ***3.3***; there are copies of the three data sheets available for each ward in every town and city of Britain. The figures in the tables are not actual values; they merely enumerate the position of actual values and are included to identify the position and number of actual values. In the top left hand corner of each table is a locational identifier; in this case the first two lines tell us that this data refers to Central ward in Birmingham and the third entry on the third line refers to a National Grid Reference of a point within the ward.

The data in ***Table 3.1*** were obtained from a 100% sample of the population in the ward. The data are arranged in boxes, numbered in the left hand corner of the box, in this case from 1 to 14. The data in boxes 1–9 are given in raw form, referring to actual values, while the data in boxes 10–14 are given for either per 100 or per 1000 persons. This standardised data allows easy calculation of percentages. Data in this 100% population table includes the number of persons in the ward, the age structure and marital status of the population, both singly and cross-classified, and the birthplace by country of the population. The data in the household table, ***Table 3.2***, were again obtained from a 100% sample of the population. Only the data in box 18 are given in the form of numbers per 1000 households; all the other boxes in this table refer to absolute values. The information available from this table includes the number of households, the housing tenure and quality of households, the type of households and the housing tenure of both households and persons cross-classified with respect to the number of rooms, the lack or sharing of amenities (bath, inside w.c. and hot water) and room density. Unlike the other two tables, the data given in ***Table 3.3*** were obtained from a 10% sample of the population. This small sample is unlikely to be a problem when using ward data but the sampling errors for enumeration district data can be considerable if very small values, e.g. less than 10, are recorded. All data in this table are in the form of absolute values. The information available in this table includes the number, socio-economic and marital status of migrant households, the socio-economic status of persons and households and the socio-economic status of persons cross-classified with reference to mode of journey to work and the type of industry.

Further information on the tables and some of the terms which are used are available in the explanatory notes which accompany the 1971 Census, small area statistics. These notes

TABLE 3.1 1971 CENSUS, SMALL AREA STATISTICS, 100% POPULATION

* BIRMINGHAM C3 * CENTRAL PART OUTSIDE C.C. 123 1234/A01 SU 123456	SMALL AREA STATISTICS (WARD LIBRARY)	100% POPULATION	CENSUS 1971

1

Total population present	
Visitors res. in L.A.	
Visitors to L.A. res. in G.B.	
Visitors to G.B.	
Total pop. in private H/H	
Res. at address of enum.	
Total pop. not in private H/H	
Res. in the estab.	

2

	NOT IN PRIVATE HOUSEHOLDS					
	All present		Resident staff		Other residents	
	Males	Females	Males	Females	Males	Females
Hotels & B/Houses						
Children's homes						
Old people's homes						
Psychiatric hosps.						
Other hospitals						
Schools & colleges						
Other non-pvte H/H						
Total						

3 RESIDENTS WITH NEITHER PARENT BORN IN U.K.

	BOTH PARENTS BORN IN NEW C.				OTHERS			
AGE	SWDM	MM	SWDF	MF	SWDM	MM	SWDF	MF
0–4								
5–14								
15–24								
25–34								
35–44								
45–59								
60–64								
65+								
Total								
Total E.A.								

4 AGES TO 21 YEARS

	Males	Females
0		
1		
2		
3		
4		
0–4		
5		
6		
7		
8		
9		
5–9		
10		
11		
12		
13		
14		
10–14		
15		
16		
17		
18		
19		
20		
21		

5 AGES OF ALL ECONOMICALLY ACTIVE · **6** PERSONS NOT IN PRIVATE HOUSEHOLDS · **7** PERSONS PRESENT IN PRIVATE HOUSEHOLDS

	5	5 FEMALES		6 MALES		6 FEMALES		7 MALES		7 FEMALES		7 Children ever born to
	Males	SWD	Married	SWD	Married	SWD	Married	SWD	Married	SWD	Married	married women
0–14												
15–												
20–												
25–												
30–												
35–												
40–												
45–												
50–												
55–												
60–												
65–												
70–												
75–												
Total												
Working												
Seek wk												
Sick												
Retired												
Other												

8 BIRTHPLACES OF PRES RESIDENTS

	Males	Females
England		
Scotland		
Wales		
N.I. etc.		
Irish Rep.		
Old C.		
New C.		
Africa		
America		
Ceylon		
India		
Pakistan		
Far East		
Rem.		
Other Eur.		
O.F.& N.S.		

9 BIRTHPLACES OF THOSE ENTERING U.K. AFTER 1960

	Males	Females
Irish Rep.		
Old C.		
New C.		
Other Eur.		
O.F.& N.S.		

10 NUMBER E.A. PER 1000 OF:

Age 15 or over	Males age 15 or over	Females age 15 or over	Married Females private H/holds

11 NUMBER PER 1000 PERS PRES IN PRIVATE H/HOLDS

Aged 0–4	Aged 5–14	Stds 15 or over	Male 65 or over	Fem 60 or over	Married Females aged 16–29	30–44	45–59

12 CHILDREN EVER BORN PER 100 MARR FEM IN PVTE H/HOLDS AGED

16–29	30–44	45–59

13 NUMBER PER 1000 PRESENT RESIDENTS

Both parents born N.C.		One Parent N.C.	Person born in		
Born GB	Born NC	Born NC	G.B.	N Ireland	Irish Rep.

14 NUMBER OF VISITORS PER 1000 PERSONS PRESENT

Res GB	Others

TABLE 3.2 1971 CENSUS, SMALL AREA STATISTICS, 100% HOUSEHOLDS

SMALL AREA STATISTICS (WARD LIBRARY) — **100% HOUSEHOLDS** — **CENSUS 1971**

15	Present	Absent	With 1 car	2 or more cars
H'holds				
Persons				

16	Occupied	Vacant	No hot water	No bath	No inside W.C.	With 3 aments.
Dwellings						
Sh'd dwellings						

17	Vacant Dw'l'g	Absent h'hld	Hotels& boarding houses
Rooms			

18 NUMBER PER 1000 PRIVATE HOUSEHOLDS PRESENT

Owner Occ	Council	Unfurn-ished	Furn-ished	Not Stated	H'holds in sh'd dw'l'gs	All aments. excl.	Other all aments.	Share or lack hot w.	Share or lack bath	Share inside W.C.	No inside W.C.	Persons per room: Over 1½	Over 1 to 1½	Over ½ to 1	½-¼	Less than ½	Rooms: 1-3	7+	One person H'holds: All h'holds	Pens. pers.	Two person households: All h'holds	One pension	Two Pensions	No car

19 PRESENT PRIVATE HOUSEHOLDS WITH FOLLOWING ROOMS

	1 or 2	3	4	5	6	7 or more	Total households	Total rooms
1 person								
2 persons								
3 "								
4 "								
5 "								
6 "								
7 or more								
Total H/holds								
Total persons								
All in non-perm. bldgs. H								
P								
In perm.bldgs. Owner Occupier H								
P								
Council.New Town. SSHA H								
P								
Private un-furnished H								
P								
Private furnished H								
P								
Not stated H								
P								
All in sh'd dwellings H								
P								
do. without excl. stove and sink H								
P								

20 HOUSEHOLD TYPE

Adults (Aged 15 or over) in Household	Children aged 0-14 in household: No child	One child: 0-4	One child: 5-14	Two or more children: All 0-4	All 5-14	Others	Persons	Persons E.A
One pensionable (male)								
" " (female)								
One other (male)								
" " (female)								
2+ all pensionable								
2+ none "								
2+ only one-not pens.								
Others								
All adult type								

21

All aments. excl.	Other all aments.	Share or lack hot water	Bath: Share	Bath: Lack	Inside W.C.: Share	Inside W.C.: Lack	Persons per room: Over 1½	Over 1 to 1½	No car

TABLE 3.3 1971 CENSUS, SMALL AREA STATISTICS, 10% SAMPLE

* BIRMINGHAM CB

* CENTRAL
PART OUTSIDE C.C.

123 1234/A01 SU 123456

SMALL AREA STATISTICS (WARD LIBRARY) 10% SAMPLE CENSUS 1971

22

Age	ONE YEAR MIGRANTS WITHIN L.A. S.W.D. Males	Married Males	S.W.D. Females	Married Females	ONE YEAR MIGRANTS INTO L.A. S.W.D. Males	Married Males	S.W.D. Females	Married Females
1–								
5–								
15–								
25–								
35–								
45–								
65+ 60+								
Total								

Age	FIVE YEAR MIGRANTS WITHIN L.A.	FIVE YEAR MIGRANTS INTO L.A.
5–		
15–		
25–		
35–		
45–		
65+ 60+		
Total		

24

Women's age	HOURS OF WORK OF WOMEN IN EMPLOYMENT: All women < 8hrs	8–30hrs	> 30hrs	Married women < 8hrs	8–30hrs	> 30hrs	Mar. women with children aged under 5 < 8hrs	8–30hrs	> 30hrs
15–									
25–									
35–									
45–									
60+									
Wkg in LA									
Wkg ex LA									

23

S.E.G.	ECONOMICALLY ACTIVE OR RETIRED BY S.E.G. OF HEAD H/holds	Persons	S.E.G. OF Persons	ECONOMICALLY ACTIVE Males	Females	Married Females	PERSONS IN EMPLOYMENT With O.N. or school cert. or 'A' level	With H.N. or degree	Occupn change in yr.	Wkg. ex LA	PERSONS ACTIVE BUT NOT IN EMP.	RET'D MALES
1												
2												
3												
4												
5												
6												
7												
8												
9												
10												
11												
12												
13												
14												
15												
16												
17												
Total												
	NEVER ACTIVE			INACTIVE			NOT IN EMP					
Students									–	–	–	–
Women < 60									–	–	–	–
Others										–	–	–
Total									–	–	–	–

25

	NUMBER OF FAMILIES WITH NUMBER OF DEPENDENT CHILDREN 0	1	2	3	4	5+
Lone parent						
Married couple						

26

S.E.G. of h/hold head	HEAD 5 YEAR MIGRANT WITHIN L.A. H/holds	Persons	HEAD 5 YEAR MIGRANT INTO L.A. H/holds	Persons	TOTAL Cars & Vans
1 2 13					
3 4					
5 6					
8 9 12 14					
7 10 15					
11					
16 17					
Total inc. N.C.					

27

Grouped S.E.G.	TRAVEL TO WORK BY Car	Bus	Train	M'Cycle	P/Cycle	Foot & none	Other & not stated
1 2 13							
3 4							
5							
6							
8 9							
7 10 15							
11							
12 14							
16 17							
Wkg in LAA							
Wkg ex LAA							
Self emp.							

28

INDUSTRY OF EMPLOYED PERSON Agri.	Mining	[illegible]	Constr	Utilities Transport	Distrib. Services	Nat & Loc Govt & Def

TABLE 3.4 DEFINITION OF TERMS USED IN 1971 CENSUS, SMALL AREA STATISTICS

New C. (N.C.)	–	New Commonwealth – all countries of the Commonwealth except New Zealand, Canada and Australia.
Household	–	group of persons living together who share common catering arrangements.
Dwelling	–	separate accommodation for one or more households. Households share a dwelling if they share any room or use the accommodation of others to obtain access to their own accommodation.
Amenities	–	hot water tap, fixed bath or shower and a w.c. within the building (within the dwelling in Scotland).
S.E.G.*	–	socio-economic group. These are:
		(1) and (2) employers and managers in central and local government, industry or commerce;
		(3) and (4) professional workers, self-employed and employees, e.g. doctor;
		(5) and (6) intermediate and junior non-manual workers, e.g. clerk in post office;
		(7) personal service workers, e.g. postman;
		(8) and (9) skilled and foremen manual workers, e.g. fitter;
		(10) semi-skilled manual workers;
		(11) unskilled manual workers;
		(12) own-account workers (other than professional), e.g. self-employed grocer;
		(13) and (14) farmers – employers and managers;
		(15) agricultural workers;
		(16) members of armed forces;
		(17) occupation inadequately described.
S.W.D. (M)(F)	–	single, widowed and divorced (males) (females).
Migrant	–	a person whose current address is different from his/her address 1 or 5 years ago.
L.A.A.	–	local authority area.
H P (box 19)	–	households and persons.
S.S.H.A. (box 19)	–	Scottish Special Housing Association.

*More detailed information is available in ***Classification of Occupation*** H.M.S.O. (1970).

are published by the Office of Population Censuses and Surveys and are recommended reading before embarking on extensive use of the data. Summary definitions of some of the terms used in these tables are given in *Table 3.4*.

Small-area statistics are not freely available and ward data for specific local authority areas have to be purchased from the Census Office. However, most local authority planning departments, some public libraries and many university libraries have the data concerning the local area. Local authorities are obliged to make this data available to the public, so if you want the ward data for your local town or city go to your local planning department and ask for permission to copy out whatever ward data you require. To provide a readily available data base, 14 variables have been abstracted from the 1971 Census relating to the 28 wards in Bristol and these variables are presented in Appendix B. The distribution of these wards is shown in the accompanying map in Appendix B. In the subsequent exercises of this chapter reference will be made to this data base.

3.2 USE OF CENSUS DATA

Even a cursory glance at the census sheets reveals the enormous range of data which is provided. In considering this potentially enormous data set it is worth recalling the comments made in the introductory chapter on the need for an explicit framework for collecting and analysing data. Given this large data set the possibilities of various computational and mapping exercises are very large. In the remainder of this chapter only a small number of possible uses are mentioned; hopefully the reader will be able to think of many more.

3.3 TESTING MODELS OF URBAN STRUCTURE

The data in the census can be used to test various models of urban structure. From the welter of empirical urban studies a composite model of urban structure can be suggested. This model, shown in *Figure 3.2*, suggests that the two dimensions of socio-economic status and stage in the life cycle are distributed in a distinct spatial pattern.

3.3.1 Socio-economic status

In 1939, after the examination of various cities in the United States, Homer Hoyt suggested that social groups were distributed in a sectoral pattern within the city. He maintained that households of different socio-economic status were located in distinct sectors radiating out

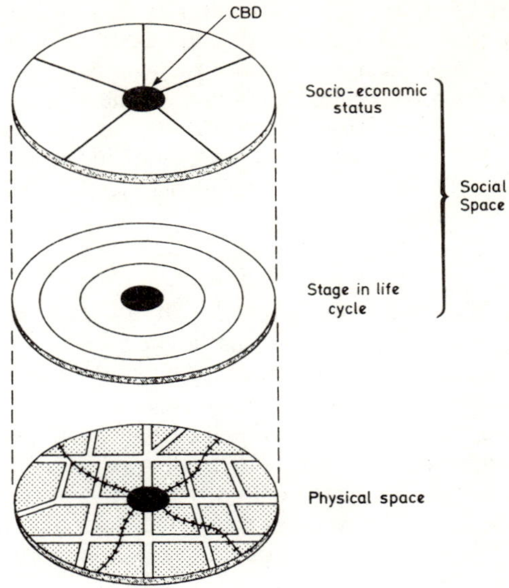

Figure 3.2 A model of urban structure

from the central business district. High status households, for example, were located in sectors running from the central business to the edge of the city away from industrial land-use areas. We can test this hypothesis by examining the percentage of persons in socio-economic groups 1–5, the highest income category, for each ward in Bristol.

As a first stage in the analysis map the wards with a percentage in socio-economic groups 1–5 higher than the Bristol average. Are these wards in a recognisable sector? Clearly there is an easily defined sector running from Cabot to Henbury, including the wards of Clifton, Redland, Bishopston, Durdham and Westbury, but the two peripheral wards of Stapleton and Stockton 'upset' the neat sector pattern. These two wards were the site of recent construction of owner-occupied dwellings. A second stage in the analysis would be to rank the wards in terms of percentage in socio-economic groups 1–5. Are the first five ranked wards in a definite sector? Does this sector extend to the boundary of the city? Similarly, we can examine the distribution of other socio-economic groups. The wards with a percentage of persons in socio-economic groups 5, 6 and 8, 9, 12 and 14 higher than the Bristol average could be mapped and ranked. Are these patterns sectoral? Is the resultant pattern similar to the picture of socio-economic status distribution shown in ***Figure 3.2***. Examine the tenure characteristics of the wards which do not fit the sector model and remember that Hoyt's model was derived from a society where local authority public housing is relatively unimportant compared to British cities.

3.3.2 Stage in life cycle

Households can be differentiated with respect to their stage in the life cycle; four stages can be identified. The ***pre-child*** household consists of young single-person households or recently married couples without children. As the name implies, the ***child-rearing*** household is composed of adult members and one or two children under 14 years of age and the ***child-launching*** household contains children between 14 and 20 years of age. The ***post-child*** households consist of more elderly persons —usually retired — with no children. Each of these stages are associated with differing housing requirements: pre-child and post-child households require less space than child-rearing households and, because of the growing children, child-launching households demand more housing. Given the distribution of housing in which smaller houses and flats are located closer to the central city, with more spacious housing located further from the centre, we can suggest a concentric pattern for the distribution of households at different stages in the life cycle. Although it is difficult to associate stages in the life cycle with distinct age categories, we can test the concentric patterns at an aggregate level by examining the demographic variables available in the Census. For example we can examine the distribution of the percentage of the population 0–14 years of age. According to the model the value for this variable should rise towards the city boundary, as shown in ***Figure 3.3***. Alternatively we can consider the distribution of the percentage of the population 15–34 years of age which we can take as the pre-child households; according to the model the values for this variable should increase closer to the city centre. More generally we could rank the wards with respect to each of these variables and note the resultant spatial distribution.

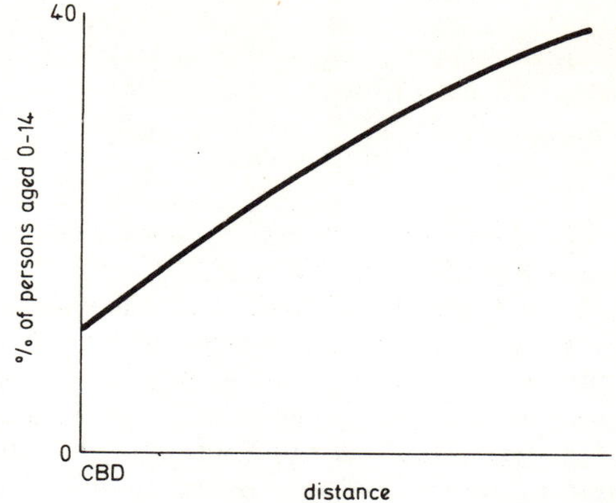

Figure 3.3 Hypothetical relationship between age and distance from city boundary

A more sophisticated testing of the proposed sectoral and concentric patterns of socio-economic status and stage in the life cycle will use the one-way analysis of variance. The first step in this procedure is to identify a grid of zones and/or sectors. If we are interested in testing the sector model we might decide to identify, say, five sectors radiating out from the city centre which together incorporate all the wards of the city and whose boundaries do not cross ward boundaries. The second step is to set up the null hypothesis that there is no significant difference between sectors in the number of persons in socio-economic group 1–5. To test this hypothesis we can use the one-way analysis of variance and estimate the F ratio. Calculate the overall mean of the number of persons in socio-economic group 1–5 for all the wards in the city. The ***total variation*** is found by adding all the squared deviations of each ward's value from this overall mean and the ***total between sector variation*** (B.S.) is found by squaring the deviation of the sector means from this overall mean and multiplying by the total number of wards. The ***total within sector variation*** (W.S.) is then found by subtracting the total between sector variation from the total variation. The F ratio is calculated in the following manner:

$$F = \frac{\text{B.S.}/(\text{nos. of sectors} - 1)}{\text{W.S.}/(\text{nos. of wards} - \text{nos. of sectors})}$$

If we use five sectors, there are 28 wards in the city and the B.S. and W.S. figures are respectively 144 and 216 then:

$$F = \frac{144/4}{216/24} = \frac{36.0}{9.0} = 4.0$$

This calculated F value can be compared with the tabulated values of F which are shown in ***Table 3.5***. For five sectors (K) and 28 wards (N) the degrees of freedom are respectively ($K - 1 =$) 4 and ($N - 1 - K =$) 22. Given these degrees of freedom the ***estimated F value*** is 2.31 and, since the actual F value of the above equation is greater than this estimated F value, then we can reject the null hypothesis and state that, at the 95% confidence level, there is a significant variation between sectors of the numbers of persons in socio-economic group 1–5. ***If the actual value of F is smaller than the estimated value of F, then the null hypothesis is accepted***. Similarly we can identify a system of zones and use the data referring to stage in the life cycle to test, by means of the analysis of variance, the concentric model. In the two-way analysis of variance, the effects of both zones and sectors can be estimated (Murdie, 1976).

TABLE 3.5 ESTIMATED *F* VALUES AT THE 95% CONFIDENCE LEVEL

Degrees of freedom \ Degrees of freedom	1	2	3	4	5
1	161.45	199.50	215.71	224.58	230.16
5	6.60	5.78	5.40	5.19	5.05
10	4.96	4.10	3.70	3.47	3.32
15	4.54	3.68	3.28	3.05	2.90
20	4.35	3.49	3.09	2.86	2.71
25	4.24	3.38	2.99	2.75	2.60
30	4.17	3.31	2.92	2.68	2.53
40	4.08	3.23	2.83	2.60	2.44
60	4.00	3.15	2.75	2.52	2.36
120	3.92	3.07	2.68	2.44	2.29
∞	3.84	2.99	2.60	2.37	2.21

It is important to note that these models of urban structure were devised in the context of North American cities where public housing constitutes less than 5% of the housing stock. The existence of a large public housing sector in Britain means that British cities do not necessarily show the patterns suggested by these models. Public housing may reinforce the zonal structure of the stage in the life cycle dimension by having child-rearing households on the periphery of the city but it may modify the sectoral pattern of the socio-economic status by placing lower-income households around the edge of the city.

3.4 DETECTING RELATIONSHIPS

The data in the Census can also be used to examine the relationships between any pair of variables. For example if we are interested in examining the relationship between housing tenure and mobility we can plot the relevant variables on a graph. The values for percentage private renters have been plotted with respect to percentage in-migrants for all the wards in Bristol in ***Figure 3.4.*** The numbers beside the points refer to the index numbers of the wards given in Appendix B. From the resultant graphs it is possible to comment on the general nature of the relationship between the two variables. In this particular case it is clear that there is a definite positive relationship between the two variables. (The term positive implies that the more private renters there are in a ward the higher will be the rate of in-migration.) It could be suggested that this relationship reflects the ease of entry into and within private renting compared to either owner-occupation or public housing and the fact that private renting is the predominant form of tenure for the younger, more mobile households. For households coming into the city, private renting provides a convenient tenure until either a house is bought or rented from the local authority or until the household moves to another city.

While the construction of a graph permits the visual display of the relationship between pairs of variables it does not allow a precise estimate of the strength of this relationship. To calculate the strength of the relationship we can use the ***Spearman Rank Correlation Coefficient*** (R_s). Each variable is ranked in order of increasing size and the difference between the values of the ranks (D) at each observation are squared (D^2) and then summed (ΣD^2). The rank correlation is then found by the following equation:

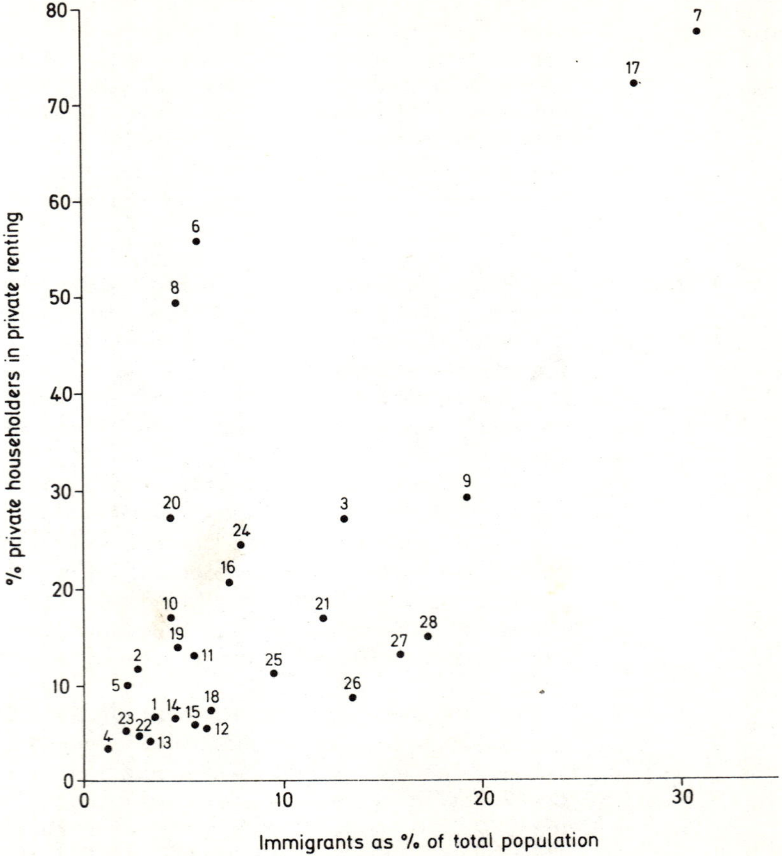

Figure 3.4 Relationship between private renting and in-migration for Bristol wards

$$R_s = 1 - \frac{6 \times \Sigma D^2}{N^3 - N}$$

where N is the number of observations. The values of R_s range from +1.0 which signifies a perfect positive correlation to −1.0 which is a perfect negative correlation. The estimation of the *Spearman Rank Correlation Coefficient* is highlighted in *Table 3.6*.

TABLE 3.6 ESTIMATION OF R_s (IMAGINARY DATA)

Ward	Percentage private renters (p.r.)	Percentage in-migrants (m)	Rank (p.r.)	Rank (m)	D	D^2
A	40	15	8	8	0	—
B	35	20	9	6	3	9
C	25	15	10	8	2	4
D	25	10	10	11	1	2
E	15	5	12	12	0	—
F	65	25	2	3	1	2
G	55	25	5	3	2	4
H	50	20	6	6	0	—
I	45	15	7	8	1	2
J	65	25	2	3	1	2
K	60	30	4	2	2	4
L	75	45	1	1	0	—

$N = 12$ $\Sigma D^2 = 29$

$$R_s = 1 - \frac{6 \times \Sigma D^2}{N^3 - N} = 1 - \frac{6 \times 29}{1728 - 12} = 0.90$$

The R_s value estimated in *Table 3.6* suggests that there is a very strong positive relationship between the two variables. Is this value significant? Could it be the result of chance, random factors? We can test the significance of the *Spearman Rank Correlation Coefficient*, when using more than 10 observations, by using the Student's t test. The t value is simply calculated by the following equation:

TABLE 3.7 CRITICAL VALUES OF STUDENT'S '*t*'

Degrees of freedom	Levels of significance 0.05	0.01
5	2.57	4.03
6	2.45	3.71
7	2.37	3.50
8	2.31	3.36
9	2.26	3.25
10	2.23	3.17
11	2.20	3.11
12	2.18	3.06
13	2.16	3.01
14	2.15	2.98
15	2.13	2.95
20	2.09	2.85
25	2.06	2.79
30	2.04	2.75
40	2.02	2.70
50	2.01	2.68
120	1.98	2.62
∞	1.96	2.58

$$t = R_s \sqrt{\frac{N-2}{1-R_s^2}}$$

Using the results from *Table 3.6* this gives

$$t = 0.9 \sqrt{\frac{12-2}{1-0.81}} = 6.53.$$

The value can be compared to the critical values of ***t*** provided in *Table 3.7*. The appropriate critical value of ***t*** is found by selecting a significance level at 10 degrees of freedom (degrees of freedom = $N - 2$). If we adopt the 0.05 level the appropriate critical value is 2.23. Since the estimated value of ***t***(6.53) is ***greater*** than the critical value, then we can say that in only 5% of cases could a relationship of this strength have occurred by chance. If we adopt the 0.01 level, the critical value is 3.17 and since the estimated value is ***greater*** than this, then we can say that in only 1% of cases could a relationship of this strength have occurred by chance. ***If the estimated value of t is smaller than the critical value of t then the strength of the relationship is not significant.***

The plotting and computation of correlation coefficients allows us to detect the relationship between pairs of variables and to estimate the strength of this relationship. There are a number of pairs of variables which are worth considering. Examine the relationship between the variables associated with housing tenure and socio-economic status; housing tenure and ethnic status; socio-economic status and housing quality; and housing quality and housing tenure. Before plotting the values and computing the R_s values, suggest both the form and the strength of the relationship you expect to find given your previous knowledge of cities and housing. Do the relationships you discover correspond with your prior expectations?

Before embarking on this exercise it is important to consider some technical and more general points.

3.4.1 Technical points

1. Before constructing the axes of the graphs examine the range of values so that the most convenient scale for the axes is chosen.
2. Experiment with a range of scales since alternative scales may highlight more clearly the form of the relationship.

3. Examine the graph and the correlation coefficients with respect to three *ideal* types. A strong positive relationship where all the points are located around a straight line as shown in ***Figure 3.5a*** occurs when high values of one variable are associated with equally high values of another variable. A strong negative relationship as shown in ***Figure 3.5b*** occurs when high values of one variable are associated with equally low values of the other variable and vice versa. No relationship is found when the points are randomly distributed as in ***Figure 3.5c.***

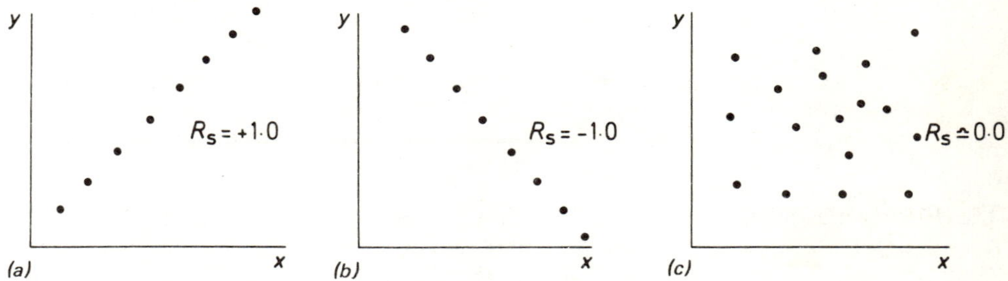

Figure 3.5 Three ideal types of statistical relationship

4. Examine the characteristics of ***outliers***. In ***Figure 3.4***, for example, there is a strong positive relationship between the two variables but wards 6 and 8 do not follow this general pattern; they are termed outliers. An examination of the spatial location and characteristics of these wards may reveal the reasons why they do not follow the general pattern.

3.4.2 General points

1. Since the relationships we are considering often refer to relationships between individual household characteristics it is important to recall the problems of the ecological fallacy. The ecological fallacy does not negate the examination of aggregate data but it does mean that our conclusions can only be tentative, not definite, assertions. The relationships found at the ward data level are only indicative of relationships found at the individual level and are best used as an exploratory guide.

2. Be careful to distinguish between statistical and causal relationships. The correlation coefficients are statistical relationships. Although we can use the resultant figures to suggest causal relationships, they cannot be used definitely to prove causal relationships.
3. In examining any two variables it is important to consider the effects of other variables on the relationships. For example, the relationship between private renting and in-migration is influenced by the demographic characteristics of the population in the wards. Younger households, who often move into private renting, are more likely to move than older households.
4. A more sophisticated treatment of variable relationships is available in the use of the linear regression model (Johnston, 1978).

3.5 URBAN HOUSING

The Census provides a rich source of data relating to housing characteristics. Housing is an important element of cities and one of the main foci of urban planning and public interest. Against this background it is important to discover the spatial structure of the housing market in cities.

3.5.1 Tenure

Although it is useful to map each housing tenure variable individually, the data can be summarised by the use of special graph paper. Since the three tenure variables — percentage owner-occupiers, percentage local authority tenants and percentage private renters — add up to 100% in each and every ward of the city this characteristic can be used with respect to three-sided graph paper. In ***Figure 3.6a*** the three tenure variables for all the wards in Bristol have been plotted and mapped; on this map the central business district (CBD) is situated in St Philiph and Jacob ward. The construction of this diagram and the correct use of three-sided graph paper is highlighted in ***Figure 3.6b***.

From ***Figure 3.6a*** and ***b*** it is clear that housing tenure types are not randomly distributed throughout the city. ***Figure 3.7*** is a general model of the spatial distribution of tenure types in British cities. Private rented accommodation is invariably found in the inner city. Before

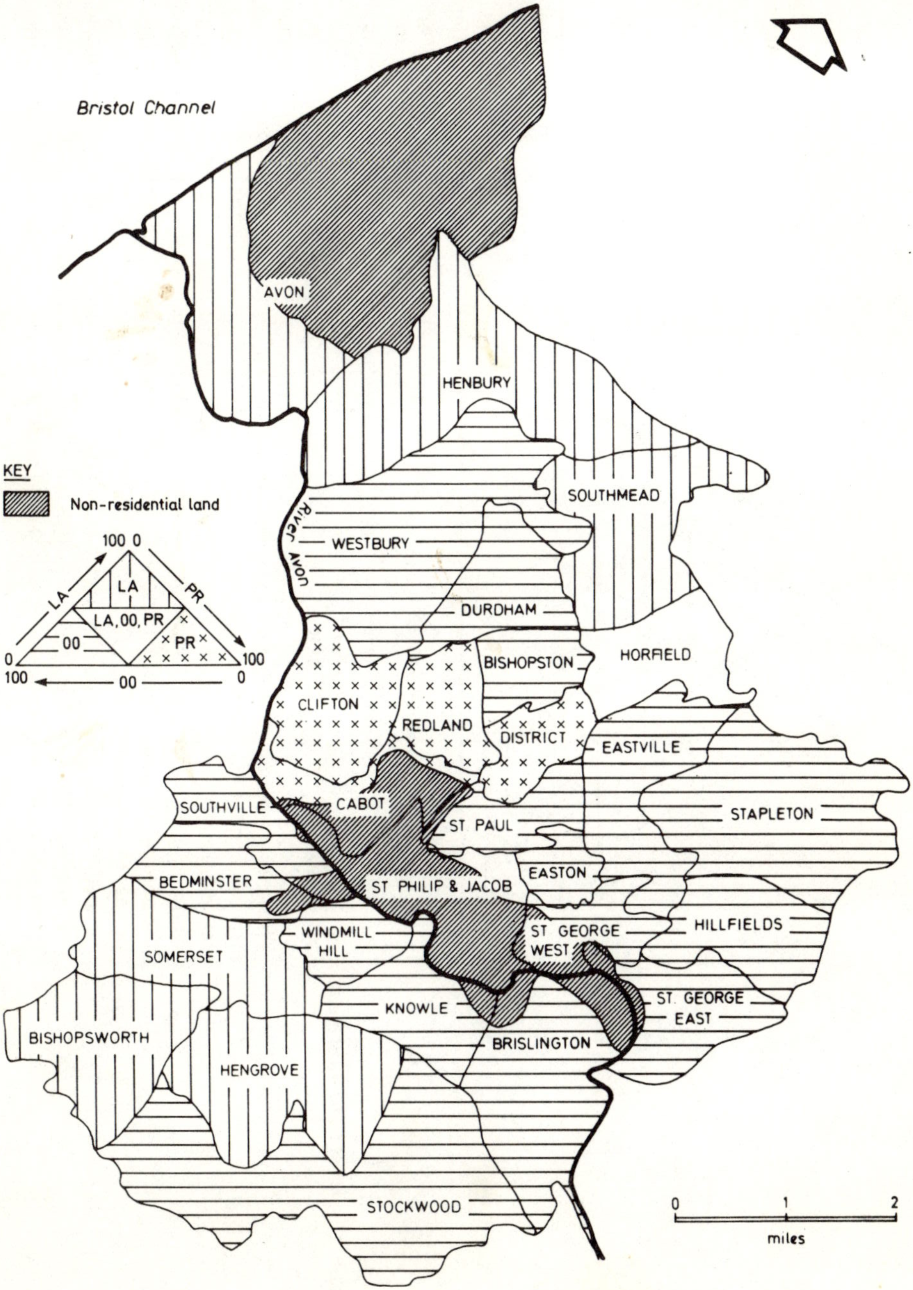

Figure 3.6a Housing tenure types in Bristol, 1971

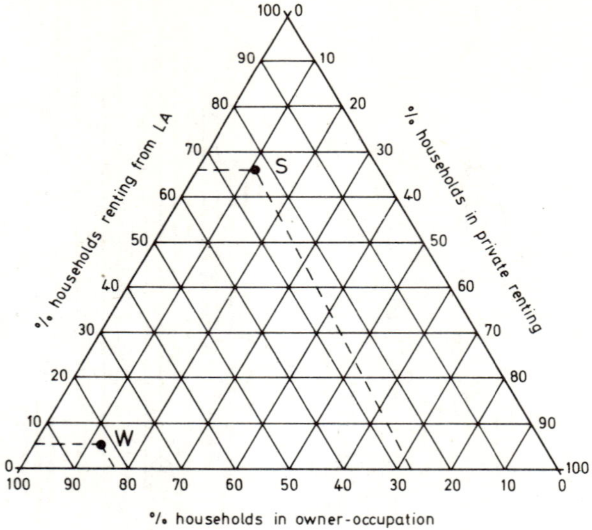

	Ward	*% households in owner-occupation*	*% households renting from local authority*
S	Southmead	28.4	66.3
W	Westbury	81.8	5.2

Figure 3.6b The use of three-sided graph paper

World War I private renting was the predominant form of tenure; in 1914 almost 90% of the population lived in privately rented accommodation. Since then the decline of private renting both in absolute and relative terms has continued (it now constitutes less than 20% of the housing stock), and because there are very few houses constructed specifically for private renting, the subsequent growth of owner-occupation and public housing has left the private renting sector with a distinct inner city distribution. Public housing was constructed on a large scale after World War II. Owing to the high number of war-damaged properties and the popular demand for good quality housing the local authorities, with financial assistance from the central government, began to build large council estates. Because of the

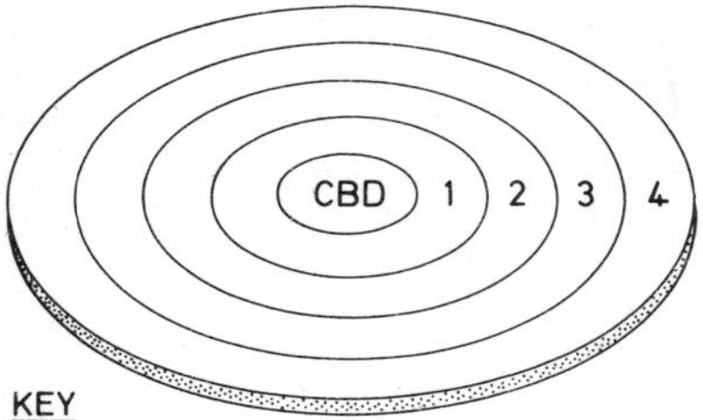

KEY
1 Private renting
2 Owner-occupation
3 Renting from L.A.
4 Owner-occupation

Figure 3.7 A general model of tenure distribution in British cities

desire for low-density housing and the lack of available space most of the housing was constructed at the edge of the city and nearly every city has a ring of council housing around the city boundary. In the sixties and early seventies as the amount of green field sites began to diminish and the local authorities were unable to build beyond their jurisdiction, some inner city council housing was constructed. Owner-occupation began to expand after World War I and now constitutes just over 50% of the national housing stock. Although there are some inner city areas of older owner-occupied dwellings, the bulk of pre-World War II owner-occupied dwellings are sandwiched between the private renting and public housing areas. Post-war owner-occupation is found in small areas within the city and, more importantly, in the suburban areas across the city boundaries. The spatial distribution of tenure types as depicted in ***Figure 3.7*** is a very generalised picture, but it does represent the basic outline and allows the deviation between the model and the reality in any one city to be quickly identified.

3.5.2 Housing quality

Recent government housing policy has been concerned with the improvement of older housing. The 1969 and 1974 Housing Acts both contain measures to improve, rather than demolish, poor quality housing. With this recent emphasis on improvement it is important to identify the poorer quality housing areas in the city. The Census provides a number of housing quality variables including the general one of the percentage of households without exclusive use of amenities and one specifically relating to high density, i.e. the percentage of households with more than 1.0 person per room. These two variables and, of course, others, can be used to identify housing problem areas. These two variables are ***direct*** indicators of housing quality. ***Indirect*** indicators of housing quality are variables which do not measure housing quality directly but can be used to identify areas of poorer housing quality. Poor housing quality is not directly associated with private renting, there are private rented dwellings which are of very high standards but, since private rented accommodation is generally older and of poorer quality than other tenure categories, the variable percentage private renters can be used as an indirect measure of housing quality.

The housing quality variables, both direct and indirect indicators, can be ranked and mapped. ***Table 3.8*** ranks the worst five wards in Bristol on the three variables previously

TABLE 3.8 RANK ORDER OF POOR QUALITY HOUSING WARDS IN BRISTOL

Rank	Direct indicators		Indirect indicator
	Percentage lack amenities	Percentage >1.0 person per room	Percentage private renting
1	St. Pauls	Hengrove	Clifton
2	District	St. Pauls	Redland
3	Redland	Bishopsworth	Cabot
4	Cabot	Southmead	District
5	Windmill Hill	District	Durdham

mentioned. In terms of lack of amenities, the worst wards are the inner city wards with St. Pauls and District having more than twice the Bristol average of households lacking amenities. On the overcrowding index, the worst wards include the inner city wards of St. Pauls and District and the public housing wards of Hengrove, Bishopsworth and Southmead. In the private renting measure, with respect to the direct indicators, District and Cabot make a third and second appearance respectively. Since the variables measure slightly different aspects of housing quality it is sometimes appropriate to construct a general index of housing quality. This can be achieved in two ways. The simplest method is to rank the wards with respect to all the variables you wish to consider. A simple example is shown in ***Table 3.9***. Ignore the figures in brackets for the moment and note that each of the four wards have been ranked with respect to three variables and the rank numbers have

TABLE 3.9 AN EXAMPLE OF COMPOSITE INDEX CONSTRUCTION

Wards	Rank variable *X*	Rank variable *Y*	Rank variable *Z*	Total	Composite rank (A)	Composite rank (B)
A	1	4 (8)	2	7 (11)	2	4
B	2	3 (6)	1	6 (9)	1	1
C	3	2 (4)	3	8 (10)	3	2
D	4	1(2)	4	9 (10)	4	2

been added. The composite rank under (A) is found by ranking the original rank number totals. In terms of poor quality housing the wards are now ranked B, A, C and D. A map of this composite rank index (A) will reveal the wards of poorest housing quality. However, you may feel that some variables are more important than others, some variables may be measuring the type of housing quality you are particularly interested in. In this case the variables can be weighted. In the example shown in ***Table 3.7*** it has been assumed that variables ***X*** and ***Z*** are less important or less relevant than variable ***Y***. To introduce this notion into our ranking scheme we can multiply all the rankings on variable ***Y*** by 2. The weighted rank and the new totals are given in brackets and the new index, composite rank index (B), is found by ranking the amended rank totals. With this new composite rank index the wards are now ranked B, C, D and A in terms of housing quality.

By experimenting with a number of variables and different types of weighting — all but one of the variable ranks can be weighted and a variety of different weights (e.g. multiplication by a factor of 2, 3, 4 or more) can be used — a variety of maps of housing quality can be constructed. Unless you adopt a highly unusual weighting system you will find that the wards of poorest quality housing are invariably found in the inner city. This exercise is not an end in itself. Relate your housing quality maps and areas to housing policies proposed for your city. Do the plans deal with the areas which you have identified as poor quality housing? If so, do you agree with the proposals?

An example of such an exercise can be found in a study of improvement grant distribution in Bristol (Bassett and Short, 1978). The purpose of this study was to ascertain if improvement grants — a sum of money given by the local authority to households for the improvement of private housing — were being allocated to the poorest quality housing areas. Six areas of inner Bristol were identified and defined in terms of percentage of households with exclusive use of all amenities. The percentage distribution of the total number of grants and of total grant expenditure in the whole of the city in the period 1974–1977 in these areas is shown in ***Table 3.10***. From this table it can be seen that the worst housing areas received the smallest proportion of both grant numbers and grant expenditure. Area 1, for example, was the worst housing area as identified by the index percentages of households with exclusive use of all amenities, but yet it only received less than 5% of all grants and less than 7% of grant expenditure. In general the worst housing areas

TABLE 3.10 DISTRIBUTION OF IMPROVEMENT GRANTS IN INNER BRISTOL, 1974–1977

Area	Percentage of households with exclusive use of all amenities	Percentage of total grants	Percentage of grant expenditure
1	40.5	4.7	6.8
2	51.7	7.9	7.1
3	55.4	4.9	9.5
4	61.9	7.2	16.2
5	69.4	19.2	16.5
6	73.2	19.4	16.0

were least affected by the distribution of improvement grants and these results suggest that the improvement grant policy is perhaps not the best policy for improving housing quality in the worst housing areas of the city.

3.6 THE RESIDENTIAL MOSAIC

3.6.1 The Burgess model

Intra-urban variations in physical space, social space and housing conditions combine to produce a patchwork quilt of different types of residential areas. In this respect, a city can be visualised as an urban mosaic of neighbourhoods. The structure of the residential mosaic has long held a fascination for urban researchers. One of the earliest attempts to model the nature of the residential mosaic was made by E.W. Burgess in 1925. In the context of Chicago in the first quarter of the twentieth century Burgess proposed a concentric model of residential areas as shown in ***Figure 3.8***. In this model a traverse from the central business district to the periphery of the city passes through the factory zone, the zone in transition, the zone of working men's homes and through the residential zone to the commuters' zone. The zone in transition is composed of poor quality rented accommodation inhabited by low income, manual workers, while the zone of working men's homes contains the better quality housing of the better paid manual workers. In the residential

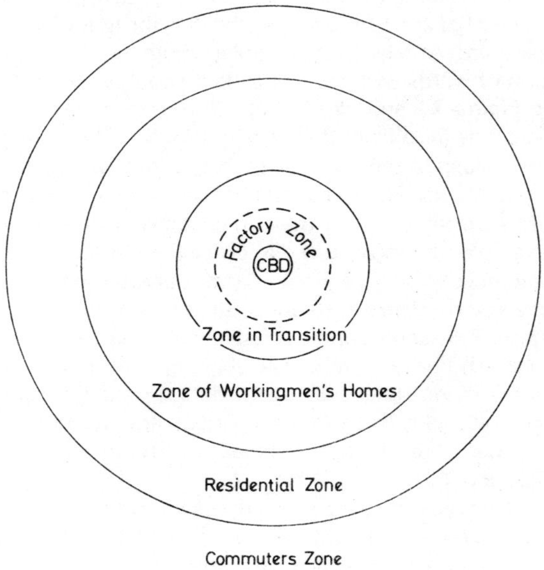

Figure 3.8 The Burgess model

and commuters' zones are found the better quality housing and the higher income non-manual workers with the best quality housing and the highest incomes being found in the commuters' zone.

Although this model is clearly a reflection of the private housing market in North America at one point in time it has exercised considerable influence on urban research in Britain. However, the large scale construction of public housing in British cities has made the Burgess model more and more irrelevant to an understanding of the structure of British cities. More recent research has been concerned less with discovering Burgess's zones and more with identifying the specific residential structure of British cities.

3.6.2 Identifying residential areas

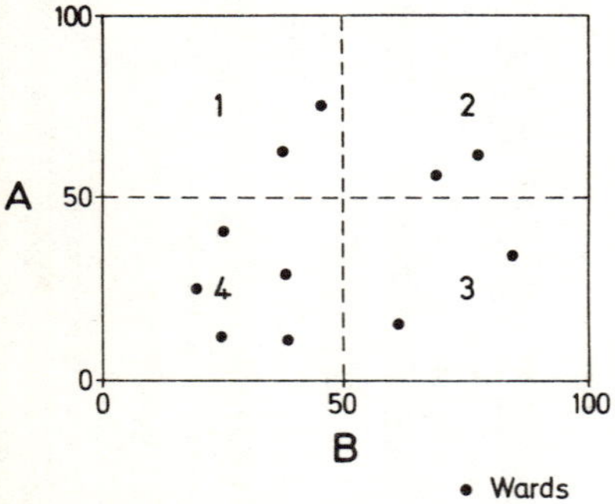

Figure 3.9 Classification of wards using two variables

We can use the data in the small area statistics of the Census to identify different types of residential areas by plotting the distribution of wards with respect to different pairs of variables. As a general example, assume that we are interested in discovering the distribution of wards with respect to two variables A and B. We can construct a diagram as shown in ***Figure 3.9*** and plot the position of each ward with respect to the two variables. We can generalise from this distribution by subdividing the diagram into four quadrants. If we then mapped the wards in each quadrant we would have a map of four types of residential areas. By considering a range of pairs of variables and by using different classification grids we can examine the rich diversity of the residential mosaic. ***Figure 3.10*** shows one example of this type of exercise where the wards of Bristol have been classified with reference to socio-economic status and tenure. Wards located in the area marked A in the key to ***Figure 3.10*** are those wards where there is a relatively high percentage of high status households and a relatively low percentage of owner-occupiers. Wards in the area marked B have a similar social profile but, in contrast, have a relatively high proportion of owner-occupiers. Using the data supplied in Appendix B select a variety of pairs of variables relating to wards in Bristol and, using different classifications, construct a number of residential area maps of Bristol and compare them with the one shown in *Figure 3.10*.

This type of exercise is important in two respects. Firstly, the resultant maps can be compared with the different models of residential structure. Secondly, the residential area classification can be used as a sampling frame for subsequent work. For example if we are interested in the levels of in-migration to the city we could construct an area classification with suitable pairs of variables to see if in-migration rates vary between residential areas. On the other hand, we might wish to discover the average house price as evinced from examination of the property section of the local newspaper. The neighbourhood typology provides a convenient sampling frame for such an exercise. Overall, the construction of residential area typologies and the resultant maps provide a convenient framework for organising and collecting data.

The classification of subareas of the city by using two variables in the type of exercise just described provides a useful introduction to describing the residential mosaic of a particular city. However, to examine this mosaic in more detail it is necessary to use more

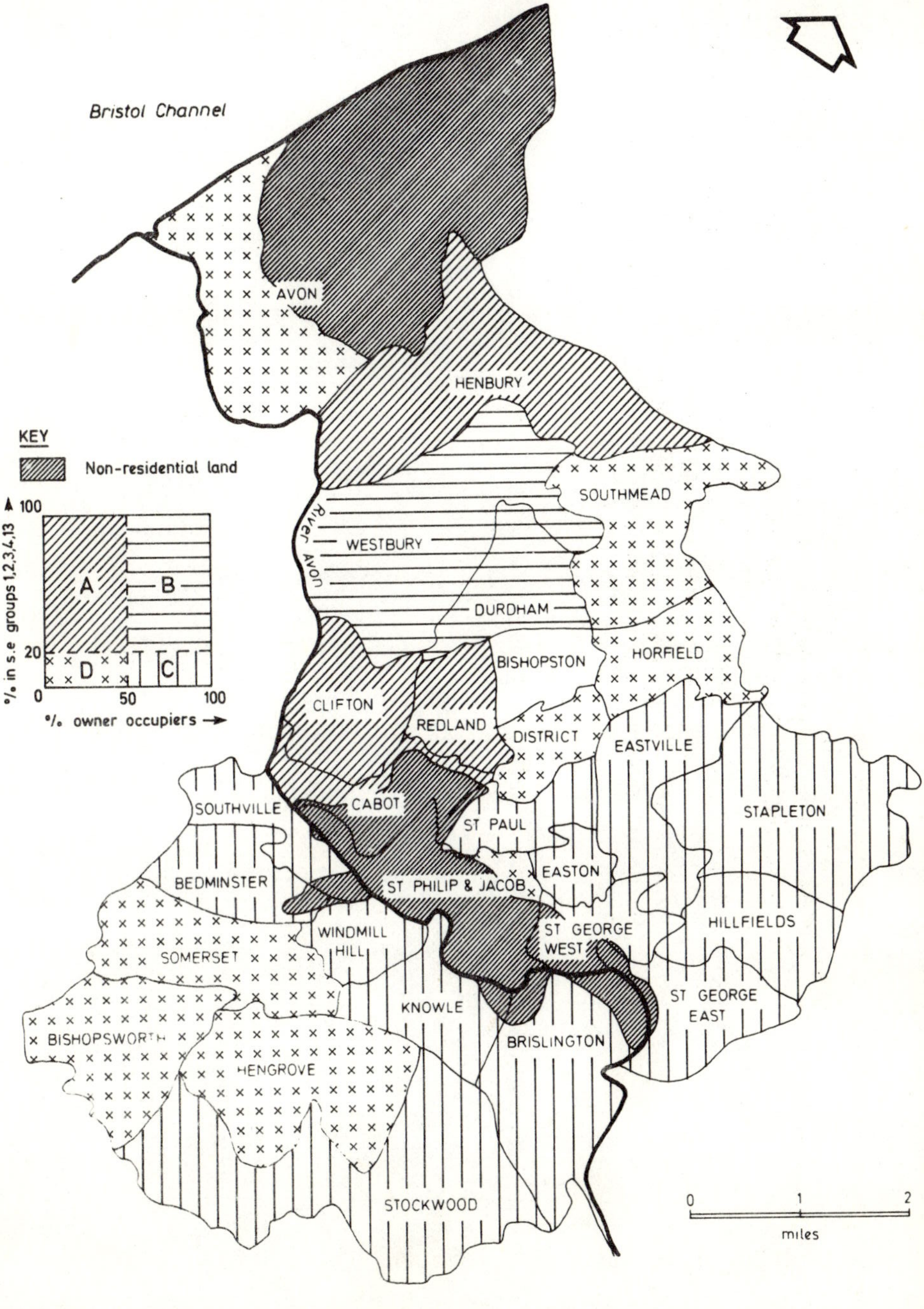

Figure 3.10 Classification of Bristol wards

than two variables for small observation units such as the enumeration districts. An analysis of this kind can be achieved by using multivariate techniques such as ***principal components analysis***. In essence, this technique generalises from a large number of variables to produce a smaller number of variables – termed components – whose variation within the city can then be mapped. This technique has been widely used to identify residential areas of the city. A general introduction to the use of principal components analysis in geography is available in Abler, Adams and Gould (1971), and Daultrey (1976), while the use of this technique in urban research is admirably covered in Johnston (1976). An example of the technique applied to enumeration districts in Bristol can be found in Short (1978a).

CHAPTER 4 RESIDENTIAL SEGREGATION

4.1 INTRODUCTION

Households of different socio-economic status are not randomly distributed throughout the city. In any major city there are recognisably 'rich' and 'poor' areas. The distribution of households of different socio-economic status and of people of different birthplaces into different areas of the city is termed ***residential segregation***. In this chapter we will discuss the data and some of the techniques which can be used to measure the degree of residential segregation.

4.2 THE DATA

The obvious source of data for measuring residential segregation in British cities are the Small Area Statistics of the 1961, 1966 and 1971 Census. For each enumeration district and ward of the city these census sources provide information on the socio-economic status and birthplace of residents (***see Table 3.3***). Before discussing the way these data can be used it is important to note their shortcomings. Firstly, they only refer to the situation at one point in time. The data in the Census refer to the characteristics of the residents who happen to be living in the enumeration district/ward on the day the Census was taken. In stable areas this is unlikely to be a problem, but in areas where there is a great deal of population movement the Census may well be out of date only a few years after it was taken. It is, therefore, important to use the most recent data to comment on the contemporary patterns of residential segregation. Secondly, the data in the Census referring to birthplace and socio-economic status refer only to a sample of the population; a 10% sample in the case of the 1971 Census, and this means that observations with small absolute values are prone to substantial sampling errors. Finally, the data refer to fairly large areas and even in the small scale enumeration district data the information refers to groups of approximately 250 households. It may be that some patterns of residential segregation are only recorded at the level of individual streets which means that they may be masked by the use of aggregate census data. We will return to this question of scale later in the chapter.

Despite these shortcomings the census data provides the most readily available data source concerning socio-economic status and birthplace. These data can be used with various indices of segregation to measure the degree and extent of residential segregation.

4.3 MEASURES OF RESIDENTIAL SEGREGATION

4.3.1 Index of dissimilarity

The index of dissimilarity (*ID*) provides a measure of the similarity in the spatial distribution of two socio-economic and/or birthplace groups. It is found by the following equation:

$$ID_{XY} = \frac{\Sigma(X_i - Y_i)}{2}$$

where X_i represents the percentage of the X population in the ith area, Y_i represents the percentage of the Y population in the ith area and ID_{XY} is the index of dissimilarity for the X and Y groups. In effect this equation tells us that to calculate the index of dissimilarity we must add up the difference between the percentage of each group in all the areas and divide by 2. ***Table 4.1*** provides an example of the calculation of this index. The *ID*

TABLE 4.1 CALCULATION OF THE INDEX OF DISSIMILARITY

Area	X Percentage persons in socio-economic groups 1 and 2	Y Percentage persons in socio-economic groups 8, 9 and 12	$(X - Y)$ Difference
1	35	10	25
2	30	15	15
3	20	35	15
4	15	40	25
TOTAL	100	100	$\Sigma(X - Y) = 80$

$$ID_{XY} = \frac{\Sigma(X_i - Y_i)}{2} = \frac{80}{2}$$

$$\therefore\ ID_{XY} = 40$$

value ranges from 0 to 100. When the *ID* value is close to 0 this implies that the two populations have similar spatial distributions. Conversely, the closer the calculated value is to 100 the more dissimilar are the two spatial distributions and hence the greater the degree of residential segregation. The *ID* value can also be interpreted as a measure of the percentage of one population who would have to move in order to reproduce the distribution of the other population. In the hypothetical case of ***Table 4.1***, 40% of the persons in socio-economic groups 8, 9 and 12 would have to move in order to have a similar percentage distribution by area of persons in socio-economic groups 1 and 2.

TABLE 4.2 INDICES OF DISSIMILARITY FOR OCCUPATIONAL GROUPS BY CENSUS TRACTS, CHICAGO, 1950 (From Duncan and Duncan, 1955)

	a	b	c	d	e	f	g	h
a. Professional workers	—	13	15	28	35	44	41	54
b. Managers, officials		—	13	28	33	41	40	52
c. Sales workers			—	27	35	42	38	54
d. Clerical workers				—	16	21	24	38
e. Craftsmen and foremen					—	17	35	35
f. Skilled manual						—	26	25
g. Service workers							—	28
h. Semi- and unskilled manual								—

Table 4.2 presents the results of an analysis of residential segregation in Chicago which used the index of dissimilarity. Note that these figures show the correspondence between social and spatial distance for different occupational categories. The two ends of the social spectrum of occupational categories, the professional workers on the one hand and the semi- and unskilled manual workers on the other, record the highest index of dissimilarity: 54% of professional workers would have to move to another area in order to match the spatial distribution of semi- and unskilled manual workers. These results show that the closer the occupational categories are in social terms, the greater the similarity in their spatial distributions.

4.3.2 Index of segregation

If we are interested in comparing the spatial distribution of a subgroup of the population to the spatial distribution of the total population then we can use the index of dissimilarity to calculate the index of segregation (*IS*), since this index measures the extent to which a specified subgroup has a similar spatial distribution to the distribution of the total population. The index of segregation is calculated by the following formula:

$$IS_{XY} = \frac{ID_{XY}}{1 - \Sigma Y_i / \Sigma X_i}$$

where, ID_{XY} is the index of dissimilarity between the total population X and the subgroup Y, ΣY_i represents the total number of the subgroup Y in the city and ΣX_i represents the total population of the city. An example of the calculation of the index of segregation is shown in *Table 4.3*. The *IS* value ranges from 0 to 100 with values closer to 100 indicating a greater degree of residential segregation.

Table 4.4 shows the results from the aforementioned study of Chicago which used the

TABLE 4.3 CALCULATION OF THE INDEX OF SEGREGATION

Area	*Y* Percentage of *Y* subgroup (absolute nos.)	*X* Percentage of total population (absolute nos.)	Difference
1	40(80)	15(150)	25
2	35(70)	25(250)	10
3	15(30)	35(350)	20
4	10(20)	35(350)	25
TOTAL	100(200)	100(1000)	80

$$ID_{XY} = \frac{\Sigma(X_i - Y_i)}{2} = \frac{80}{2} = 40$$

$$IS_{XY} = \frac{ID_{XY}}{1 - \Sigma Y_i/\Sigma X_i} = \frac{40}{1 - 200/1000}$$

$$IS_{XY} = 50.0$$

TABLE 4.4 INDICES OF SEGREGATION FOR OCCUPATIONAL GROUPS BY CENSUS TRACTS, CHICAGO, 1950 (From Duncan and Duncan, 1955)

Major occupation category	Index of segregation
Professional workers	30
Managers, officials	29
Sales workers	29
Clerical workers	13
Craftsmen and foremen	19
Skilled manual	22
Service workers	24
Semi- and unskilled	35

index of segregation. From these results Duncan and Duncan (1955) argued that, when occupational categories are listed in order of income and prestige, the indices of segregation form a U-shaped pattern with those groups at either end of the income-prestige continuum recording the highest ***IS*** values and exhibiting the greatest degree of residential segregation. This conclusion from Chicago can be used as a hypothesis to test the residential segregation of socio-economic groups in British cities. In other words, do the indices of segregation shown in ***Figure 4.1*** correspond to the pattern in any British city?

The indices of dissimilarity and segregation can also be used to measure the degree of residential segregation by birthplace of residents. For example the results in ***Table 4.5*** are taken from a study of residential segregation in Birmingham (Woods, 1976). The initial data were taken from the Small Area Statistics of the 1961 and 1971 Censuses relating to wards in Birmingham. The indices of dissimilarity and segregation were then computed for various birthplace groups for both 1961 and 1971. The results show that, whilst there is marked segregation, especially for residents born in India, Africa and the West Indies, the pattern of segregation has changed over the 10-year period. The population born in the West Indies and Ireland show less segregation in 1971 compared to 1961, while those born in India, Pakistan, Sri Lanka and Africa have become more segregated from the rest of the population.

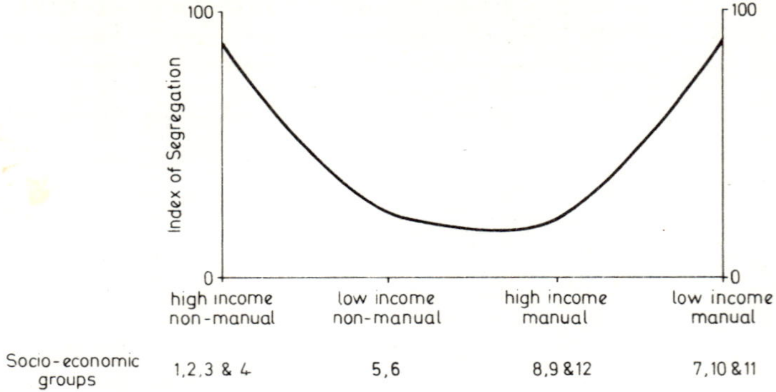

Figure 4.1 Hypothetical pattern of segregation indices for British cities

TABLE 4.5 INDICES OF DISSIMILARITY AND SEGREGATION FOR BIRTHPLACE GROUPS BY WARDS, BIRMINGHAM, 1961 AND 1971 (From Woods, 1976)

			1971		
	1	**2**	**3**	**4**	**5**
1 Total population	–	18.1*	54.3*	57.3*	49.8*
2 Ireland	27.3	–	41.7*	43.1*	35.2*
3 British Africa	41.1	37.2*	–	32.8*	37.8
4 India, Pakistan and Sri Lanka	42.1	25.3*	28.4*	–	31.4
5 British West Indies	53.8	35.9	34.5	24.9	–
			1961		

*Indices of dissimilarity: all other values refer to indices of segregation

4.3.3 Location quotient

While the indices of dissimilarity and segregation provide a measure of the degree of association between the spatial distribution of different groups of the population, they do not allow us to identify the areas in which certain groups of the population are concentrated. The degree of concentration of particular groups is measured by the location quotient.

TABLE 4.6 CALCULATION OF THE LOCATION QUOTIENT

Area		Percentage born India, Pakistan and Sri Lanka	Percentage total population	Location quotient
1		35	25	[35/25] 1.40
2		40	25	[40/25] 1.60
3		20	25	[20/25] 0.80
4		5	25	[5/25] 0.20
	TOTAL	100	100	

The location quotient – a widely used index in human geography – is calculated by dividing the percentage of the subgroup of the population in an area by the percentage of the total population in the same area. An example of the calculation of the location quotient is shown in ***Table 4.6***.If an area has an LQ value of less than 1 this indicates that the subgroup of the population under investigation is under-represented in that particular area. If an area has an LQ value greater than 1 then that subgroup of the population is over-represented in that particular area. The larger the LQ value of an area, the greater the concentration of the subgroup under investigation in that particular area. In the hypothetical case considered in ***Table 4.6*** areas 1 and 2 are over-represented in terms of the proportion of the population born in India, Pakistan and Sri Lanka, while areas 3 and 4 are under-represented; the greatest concentration of the Asian population is found in area 2.

An example of the use of the location quotient is shown in ***Figure 4.2*** which shows the distribution of LQ values calculated for the West Indian born population in the wards of London (Lee, 1973). This map reveals that the West Indian born population was very concentrated in the inner areas of London in 1966. This map was derived from data in the 1966 Sample Census. Use the data in the 1971 Census to perform a similar analysis. What is the pattern of LQ values for the West Indian born population in wards of London for 1971? Does it differ from the pattern shown in ***Figure 4.2***?

Since the three indices provide measures of different aspects of residential segregation, this suggests that any study of residential segregation should therefore use all three

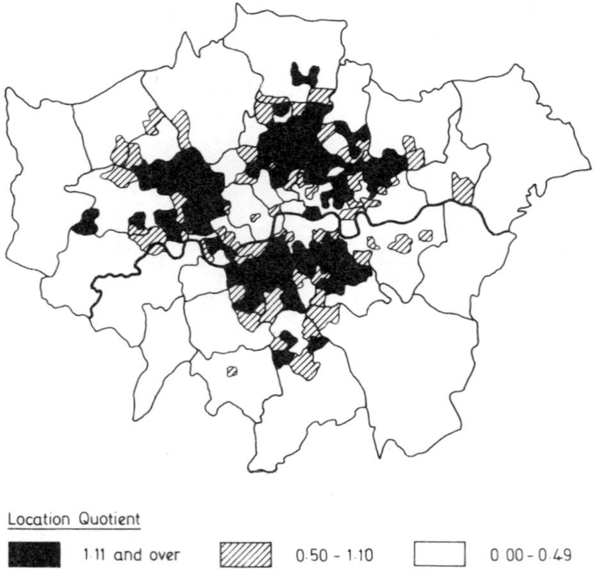

Figure 4.2 Distribution of West Indian-born population in London by location quotient

indices. The index of dissimilarity allows the distribution of a subgroup of the population to be compared to the distribution of other subgroups, the index of segregation facilitates the comparison of a subgroup distribution with that of the whole population, while the location quotient highlights the areas of concentration of a particular subgroup. If at all possible the calculation of these indices should be computed for a number of time periods, as in ***Table 4.5***, in order to highlight the evolution of segregation patterns.

4.4 THE PROBLEM OF SCALE

It must be remembered that the results from the use of these indices are dependent on the size of the observation unit. The effect of different area units is graphically illustrated in

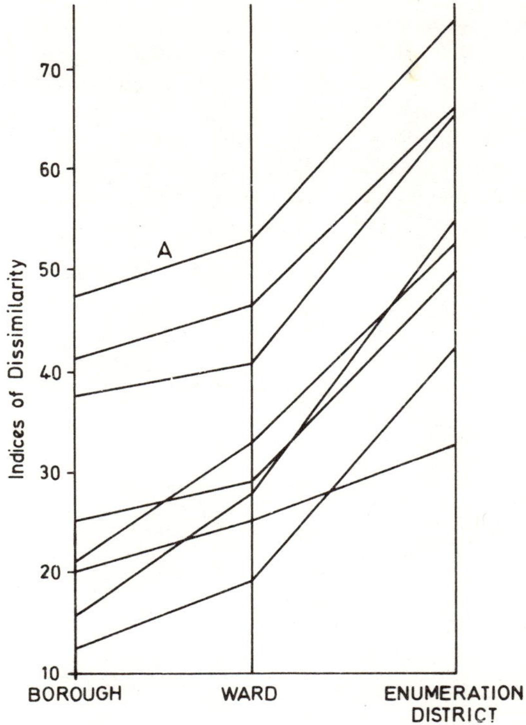

Figure 4.3 Effects of scale on indices of dissimilarity

Figure 4.3. This diagram plots the indices of dissimilarity for various birthplace groups in London in 1971 across the three spatial scales of borough, ward and enumeration district. Consider the case of the line marked A in ***Figure 4.3*** which shows that, for one particular borough, the index of dissimilarity for birthplace groups is just under 50. The mean figure for all the wards within this borough is just over 50, while the mean figure for all of the enumeration districts within these wards is well over 70. Note that when the smaller scale enumeration districts are used, then higher values of the index of dissimilarity are recorded. Also note that the lines cross as the spatial scale changes, demonstrating that the scale of analysis can influence the relative position of areas with respect to the degree of residential segregation.

The same feature is also apparent when the index of segregation is calculated. A study of Asians in Huddersfield (Jones and McEvoy, 1978) has shown that the index of segregation calculated for Asians with respect to the total population of Huddersfield varies considerably according to the scale of the analysis, as shown in ***Figure 4.4***.

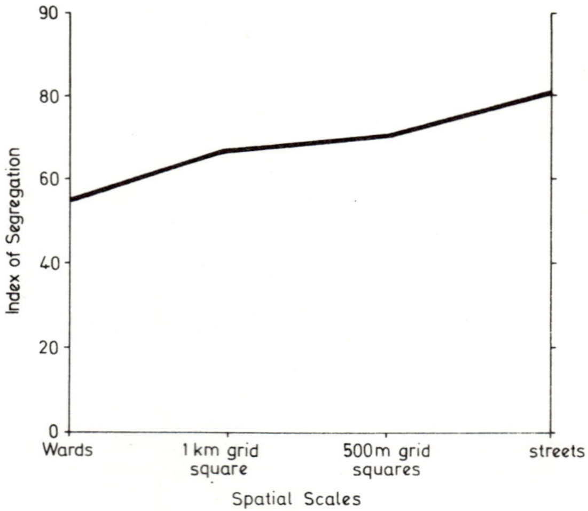

Figure 4.4 Effects of scale on indices of segregation (From Jones and McEvoy, 1978)

These two examples clearly demonstrate the importance of the scale of analysis. Large observation units may mask patterns of segregation which occur at the level of enumeration districts or indeed at the level of individual streets. Results, and conclusions, are scale specific and great care has to be taken in the form of conclusions reached from a study of segregation at only one scale of analysis. A comprehensive study of residential segregation implies not only the use of different indices for different time periods also the use of a variety of spatial scales.

4.5 THE PROBLEM OF EXPLANATION

Although the three indices which we have discussed permit the ***measurement*** of residential segregation they do not ***explain*** why residential segregation occurs or indeed how it is maintained. The calculation of the various indices provide a measure of the ***pattern*** of segregation, they do not deal with the ***processes*** which create, maintain or change these patterns. Residential segregation is created, maintained and changed by the complex interaction between a number of forces. In this section we will only touch upon a number of themes which have been more fully elaborated elsewhere (Johnston, 1971; Robson, 1975).

4.5.1 Local authority housing

Almost 30% of all housing in Britain is owned and allocated by local authorities. Since most local authority housing is allocated to those households who are unable to afford owner-occupation and, because such lower income households have heads of households predominantly in the manual occupational categories, this means that proportionately more manual households are local authority tenants than non-manual households. This characteristic of local authority housing is clearly demonstrated in ***Table 7.3*** (***see*** page 148). Since local authority housing is concentrated in distinct residential areas of the city this leads to residential segregation by socio-economic status, with local authority housing areas having a larger proportion of manual workers than some private housing areas. A high index of residential segregation for any British city in terms of socio-economic status is, therefore, partly a reflection of the way local authority housing is allocated to different types of socio-economic status categories.

4.5.2 Constraints and choices in the private housing market

In the private housing market residential segregation can be seen as the result of the differing house buying/renting powers of different socio-economic groups. ***Figure 4.5*** presents a very simple case of a city divided into three areas of different house prices and the city

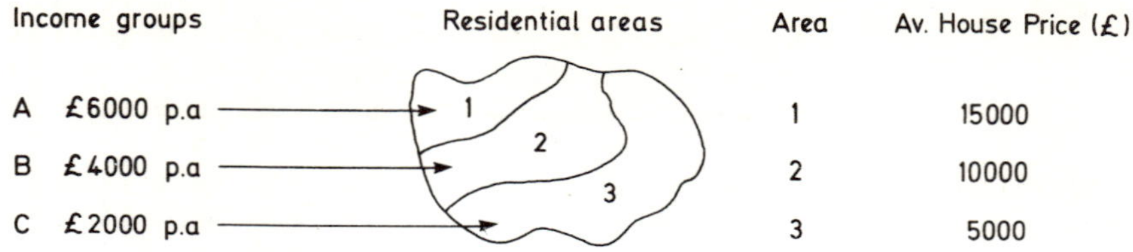

Figure 4.5 Income groups, housing and residential segregation

population divided into three income categories. Group A with the highest incomes can afford to buy housing in area 1, but the housing in this area is beyond the purchasing power of groups B and C. (In this case we are assuming that households can buy housing worth 2½ times their annual income. This is a reasonable assumption to make since most Building Societies in this country only advance mortgages – which enable households to buy housing – up to the level of 2½ times annual income.) In simple terms Groups B and C cannot afford to buy the more expensive housing in area 1. However, group B can afford to buy the houses in area 2, but the price of houses in this area is beyond the purchasing power of group C who can only afford to buy the housing in area 3. If all households bought housing commensurate with income then we would expect to find marked patterns of residential segregation between groups A, B and C in terms of areas 1, 2 and 3. This simple example suggests that, if on the one hand there is a housing market composed of housing in different areas which has different prices and on the other hand a population divided into different income levels, then we would expect to find patterns of residential segregation. This is precisely the situation in the private housing market of British cities.

Residential segregation is also a reflection of the choices exercised by some households in terms of residential location. The higher the income and status of a household the greater their degree of choice in deciding which areas of the city to live in. Those with the lowest incomes have the least choice. If we return to ***Figure 4.5***, group A can afford to live in any of the three areas 1, 2 and 3, but group C can only afford to live in area 3. Although the higher income groups can afford to live in almost any area of the city, many higher income households prefer to live in the more expensive housing areas of the city. Such a location ensures that their neighbours have roughly similar values, backgrounds and modes of behaviour. And for many households a house is not only a shelter from the elements it is also a visible reflection and indication of their standing in the community. To live in an expensive house in a select residential area is a badge of status in most circles. Thus, although higher income households have greater choice on where to live in the city, a variety of factors lead to their location in the more expensive housing areas and such patterns of residential location maintain the patterns of residential segregation.

4.5.3 Segregation by country of birth

From the results presented in ***Table 4.5*** and ***Figure 4.2*** it is clear that there is also segregation by country of birth in British cities. The most segregated groups are those who were born in Africa, India, Pakistan and the West Indies. The concentration of such households in the inner areas of major cities is a common feature of contemporary Britain. Black and brown skin-coloured households face obvious constraints in the housing market. Firstly, since many of these households have low incomes, they are only able to purchase the cheaper housing in the inner city. Secondly, the length of residence qualifications (requiring a 5–10-year period of residence in the city), which until recently was a common feature of the entry requirements for local authority housing, effectively acted as a barrier to the entry of recent immigrants to the local authority housing sector. Thirdly, the racial discrimination practised by some estate agents and building society managers meant that many coloured households could not obtain housing in areas outside the inner city.

The continued concentration of immigrant groups in certain parts of the city is also partly a function of what we may loosely term, 'choice'. In the case of immigrants, for whom British society is alien in term of customs and values, a location amongst fellow countrymen provides a familiar ***milieu*** in a strange society, a buffer against strange customs and a defence mechanism against perceived or actual threat. If immigration continues, the local community composed of a particular nationality will attract newly arrived immigrants since it provides a point of entry into an unfamiliar society; in this respect the ties of friendship and kinship with fellow countrymen may be instrumental in finding a job and a place to live, and a platform from which the immigrants can make their way in the new society. The greater the difference between the country of origin and the host country the greater the degree of clustering by nationality and hence the greater the degree of residential segregation by country of birth.

4.5.4 Conclusions

In this chapter we have shown how the Census data can be used to measure the degree of residential segregation between different groups in the city. The various indices of residential segregation provide a measure of the extent and distribution of segregation. However, it is important to stress that these indices only measure segregation — they do not explain it. The calculation of the various indices is not the end of an analysis of residential segregation, merely the beginning.

CHAPTER 5 MOBILITY AND THE CITY

5.1 INTRODUCTION

The populations of cities are constantly on the move. The changing pattern of job opportunities causes the movement of people from and to certain cities, the separation between home and workplace creates the diurnal movement of the working population within cities, while the changing space requirements associated with changes in the family life cycle generate the residential relocation of households within cities. For the sociologist Burgess, movement was an index of the metabolism of the city:

> 'Movement may be thought of as the, "pulse of the community". Like the pulse of the human body it is a process which reflects and is indicative of all the changes which are taking place in the community, and which is susceptible of analysis into elements which may be stated numerically' (Burgess, 1925).

In this chapter we will be concerned with these numerical elements. The focus will be on the range of data available in the Census, the electoral register and other sources.

5.2 MOVEMENT TO CITIES

Cities are the destination point for various kinds of population movement. Movement from rural areas to cities is termed 'rural–urban migration' while movement from city to city is termed, 'inter-urban migration'. Rural–urban migration was the predominant form of population movement in this country during the Industrial Revolution, but since then it has declined in importance. However, this form of migration is the main form of population movement in underdeveloped countries. In contemporary Britain the most important form of long-distance movement is inter-urban migration.

The existence of population movement to towns and cities poses three inter-related questions. What type of people move into particular types of cities? What is the place of origin of the people who move into certain types of cities? Where in the city do these migrants locate themselves? We can consider the data sources which allow us to answer each of these questions.

5.2.1 Who moves?

Information on the type of persons moving into towns and cities in Britain can be obtained in the 1971 Census, Migration Reports. These Reports, which are based on a 10% sample of the population, are published for each of the ten regions used by the Census. (The ten regions are North, Yorkshire and Humberside, North West, East Midlands, West Midlands, East Anglia, South East, South West, Wales and Scotland. A map of all the regional boundaries can be found at the beginning of each of the Regional Reports.) For each county, county borough and urban area with a population greater than 50 000 within the region the Reports provide information on migrants moving within the 5-year period prior to the date of the Census. Part 1 of the Regional Reports provide information on the sex, age and marital status of migrants and Part 2 contains similar information in addition to information relating to the socio-economic status of migrants.

TABLE 5.1 AGE DISTRIBUTION OF IN-MIGRANTS AND OF TOTAL POPULATION IN BRISTOL, 1971

Age categories	Migration from rest of GB to Bristol (1966–1971)		Total population of Bristol (%)
	Number	%	
5–19	752	21.2	22.2
20–34	1721	48.7	20.5
35–49	575	16.2	17.2
50–64	289	8.1	18.5
+64	128	5.8	14.4
TOTAL	3533	100.0	92.8*

*Population aged 0–14 years in 1971 are excluded

The information available in these Reports can be used to discover the characteristics of migrants moving into certain types of cities. In *Table 5.1*, for example, the information in the South West Migration Report Part 1 has been used to identify the age distribution of migrants moving into Bristol from the rest of Great Britain in the period 1966–1971. This table also provides information on the age distribution of the total population in Bristol in 1971. To aid comparison the respective figures are presented in diagrammatic form in ***Figure 5.1***. An examination of this diagram reveals that the migrants have a different age distribution to that of the total population of Bristol. The majority of migrants are in the 20–34 years of age category, this is termed the ***modal category***, and in relation to the total population of Bristol proportionately less migrants are in the 50–64 and greater than 64 years of age categories. This finding reflects the general tendency for migration to be an age-selective process. In other words ***migrants tend to be younger than the rest of the population***.

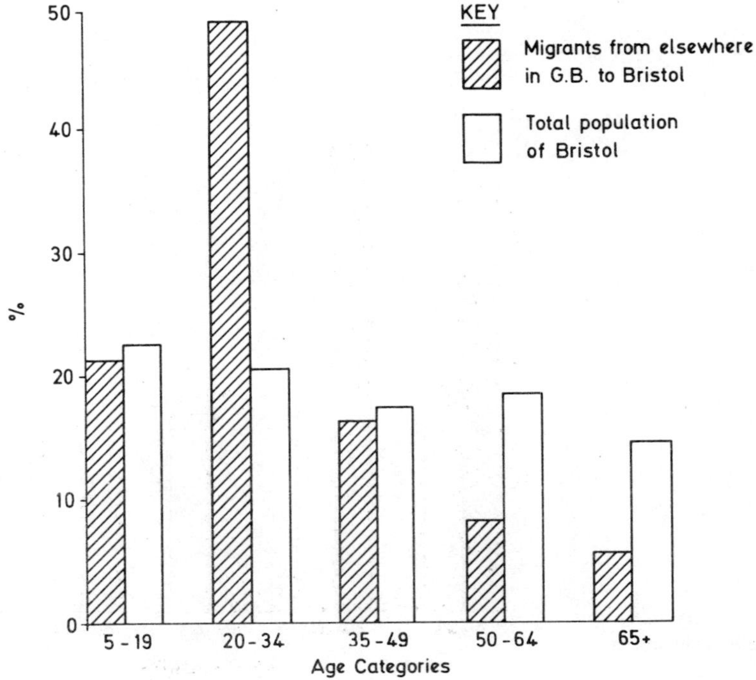

Figure 5.1 Age distribution of in-migrants and total population of Bristol

TABLE 5.2 AGE DISTRIBUTION OF MIGRANTS FROM REST OF GREAT BRITAIN MOVING INTO SWINDON AND TORBAY

Age categories	Swindon	Torbay
5–19	28.0	16.9
20–34	38.3	21.7
35–49	17.7	16.2
50–64	11.4	24.2
+65	4.6	21.0
TOTAL	100.0	100.0

This form of age selectivity is not found in all cases of migration to towns and cities. There is considerable movement of more elderly households to the retirement areas such as holiday resorts and coastal towns. *Table 5.2* provides a comparison of the age distribution of migrants from the rest of Great Britain moving within the period 1966–1971 into Torbay and Swindon. Torbay, a county borough including Torquay, is a coastal resort in Devon, and Swindon is a manufacturing centre in Wiltshire. The difference between the age distribution of the respective migrants is clearly shown in *Figure 5.2*. The modal age

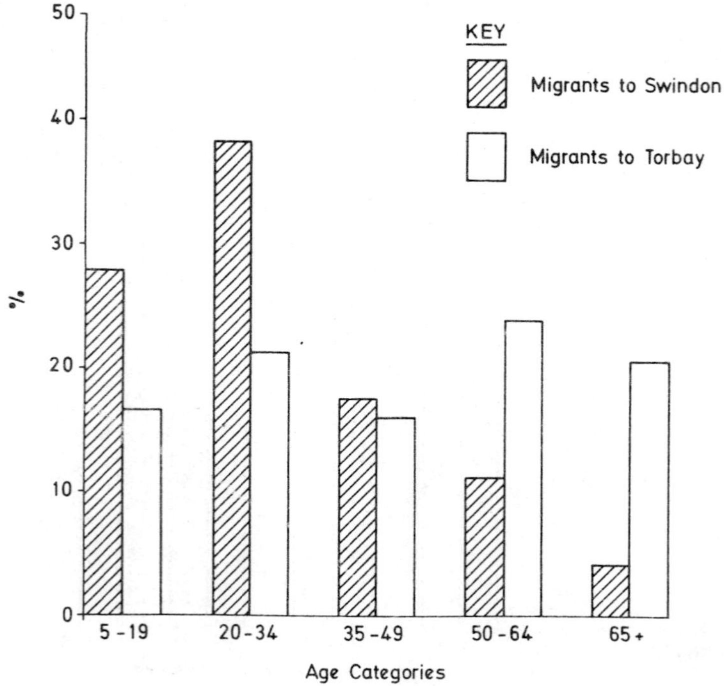

Figure 5.2 Age distribution of migrants to Swindon and Torbay

category for migrants into Swindon is 20–34 years while the modal category for Torbay is 50–64 years. While the proportion of migrants into Swindon tends to decline with age, the reverse case is found in Torbay. From these results we can suggest that Torbay is the destination for post-retirement movement. This example shows how the data available in the Regional Reports can be used to identify differences between the type of migrants moving into particular categories of towns and cities. Identify two manufacturing towns and two resort towns. Are there marked differences in the modal age category of the migrants moving into these two types of settlements?

TABLE 5.3 SOCIO-ECONOMIC STATUS OF MIGRANT AND TOTAL POPULATION OF BRISTOL

Socio-economic categories (socio-economic number as used in census)	Migrants	Total population
High income, non-manual (1,2,3,4 and 13)	39.2	15.4
Low income, non-manual (5,6)	24.7	19.6
High income, manual (8,9,12 and 14)	24.4	38.4
Low income, manual (7,10,11 and 15)	9.5	24.0
Miscellaneous (16,17)	2.2	2.6
TOTAL	100.0	100.0

Part 2 of the Migration Regional Reports provides information on the socio-economic status of migrants. This source has been used to compare the socio-economic status of migrants moving into Bristol between 1966 and 1971, with the status of the rest of the population in Bristol as shown in ***Table 5.3***. Although strict comparisons are impossible, since the migration reports use the occupation of the head of wholly moving families to determine socio-economic status while the census figures used to calculate the status of the rest of the population are based on the occupation of economically active males, broad differences can still be identified. To highlight these differences draw a diagram similar to ***Figures 5.1*** and ***5.2*** with the data given in ***Table 5.3***. It is clear from the figures in ***Table 5.3*** that non-manual workers are over-represented in migration, compared to the total population of Bristol and similarly, that manual workers are under-represented. These results suggest that migration is a selective process not only with regard to age but also with regard to socio-economic status. A greater proportion of migrants tend to have non-manual occupations than the rest of the population. To move to a different part of the country implies the necessary income to afford the move and suitable housing and employment opportunities. The socio-economic status selectivity of migration would suggest that, in relation to manual workers, non-manual workers have better housing and employment opportunities which enable them to move more easily than manual workers.

5.2.2 Origin of movement

Statistics relating to the area of former residence of migrants can be found in the 1971 Census, Great Britain Migration Tables. Comparable information is also available in the Sample Census of 1966. These Migration Tables provide a useful source of information relating to movement from and to, the 10 regions of the country, the subdivisions of regions and the conurbations. The data concerning the movement between conurbations have been used in a number of studies. Masser (1970), for example, has used this data to calibrate various versions of the gravity model.

In two respects these data are inadequate. Firstly, they do not allow the identification of the origin area of movement below the scale of conurbations and subdivisions of regions. Secondly, they do not allow us to differentiate between movement within regions (intra-regional) and movement between regions (inter-regional). In the census a move is classified as inter-regional if a regional boundary is crossed in the process of moving. However, regional and subregional boundaries may cut across functional urban areas. A move from a central city to a suburban area would, therefore, be classified as an inter-regional move if a regional boundary had been crossed.

A source of data which overcomes this latter problem is found in a paper written by Fielding (1971). Fielding divided England and Wales into 20 city regions on the basis of employed populations and journey-to-work statistics. He then persuaded the Department of Economic Affairs, as it then was, to supply migration data for the period 1960–1961. The number of moves between and within each of these city regions, disaggregated into the three age groups of 1–14, 15–59 and +60 years, are presented in an appendix to his paper. Although this source does not include data on movement since 1961 the form in which the data are presented provide a very good source of 'raw material' for a number of statistical and cartographic exercises. Fielding, for example, used the data to draw isolines of population movement into the different city regions while Cordey-Hayes and Gleave (1973) used the data for more advanced statistical analysis.

5.2.3 Destination of movement

The destination of migrants moving to the city can be ascertained from the ward and enumeration district data provided in the 1971 Census, Small Area Statistics. For each enumeration district and ward of a city this source gives the number, age, marital and socio-economic status of migrants moving from outside the local authority in the periods 1966–1971 and 1970–1971 (***see Table 3.3***). Since the results of the previous section would suggest that such migrants tend to be younger and have a higher socio-economic status than the total population, we would expect the majority of migrants to locate in wards which had a higher proportion of young people, private renting and owner-occupation than the city average. Test this proposition by plotting on a graph the relationship between socio-economic status and the number of inter-regional migrants (expressed as a percentage of the total ward population), the relationship between age structure of the ward and inter-regional migration and the relationship between the various tenure categories and inter-regional migration for all the wards of a particular city. Do the patterns in the city you have chosen confirm the proposition?

5.3 INTRA-URBAN MOVEMENT

The movement of households within towns and cities, defined as intra-urban movement, is the predominant form of residential relocation in Britain and most other developed countries. Approximately two-thirds of all households who move to another dwelling in Britain do so within the same town or city.

5.3.1 Who moves?

Information on the type of migrants moving within local authority areas in the period 1966–1971 can be obtained in the 1971 Census, Migration Reports which are published for each of the 10 regions used by the Census. Similar information is available for intra-urban migrants as for the previously discussed inter-regional migrants. For example, between 1966 and 1971, 8280 persons moved within Bristol and this figure is over twice the number of persons moving into Bristol from the rest of Great Britain in the same time period (***see Table 5.1***). The percentage distribution of these migrants by various age categories is shown in comparison to the total population of Bristol in ***Figure 5.3***. Again we can note the pronounced age-selectivity of movement. Intra-urban migrants tend to be younger than the rest of the population.

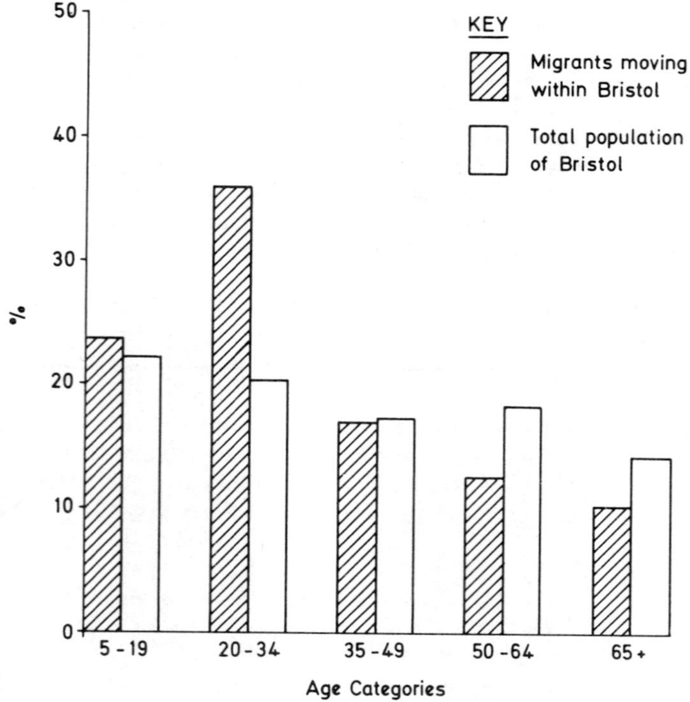

Figure 5.3 Age distribution of intra-urban migrants in Bristol

These Regional Reports can also be used to compare the age and socio-economic distribution of migrants moving within different types of towns and cities in a similar fashion to the analysis of inter-regional migrants moving into Swindon and Torbay which was discussed in the previous section.

5.3.2 Origin of movement

There is no official data source which allows us to identify the origin area of intra-urban migrants. This has prompted a number of researchers to conduct questionnaire studies with recently moved households in order to discover areas of previous residence. Short (1976), for example, has examined the pattern of movement traced by intra-urban migrants moving into four areas of Bristol. A sample of these moves into the Bedminster and St. George East wards of the city are shown in ***Figures 5.4*** and ***5.5***. These two figures highlight the short-distance nature of intra-urban movement. The majority of intra-urban migrants move over short distances. Over 48% of the movers into Bedminster and almost 32% of those moving into St. George East had a previous residence less than a mile from their present residence.

5.3.3 Destination of movement

The numbers and characteristics of intra-urban migrants moving into different enumeration districts and wards of a city can be found in the 1966 and 1971 Census, Small Area Statistics (***see Table 3.3***). Herbert (1973) has used the 1966 Census to calculate rates of intra-urban mobility for all the enumeration districts in Swansea. The rates were calculated by dividing the number of intra-urban migrants in an enumeration district by the total population of the enumeration district and then multiplying by 100 to express the result as a percentage figure. A map of the rates of mobility is shown in ***Figure 5.6***. Herbert found high rates of mobility in areas of recent house construction, high status areas and some local authority housing areas. Low rates of mobility were found in the older housing, terraced areas of the city.

A common feature of studies of intra-urban migration has been the use of the number of intra-urban migrants, available in the Census, as a percentage of the total population in an area and consideration of this variable as an index of ***population turnover***. However,

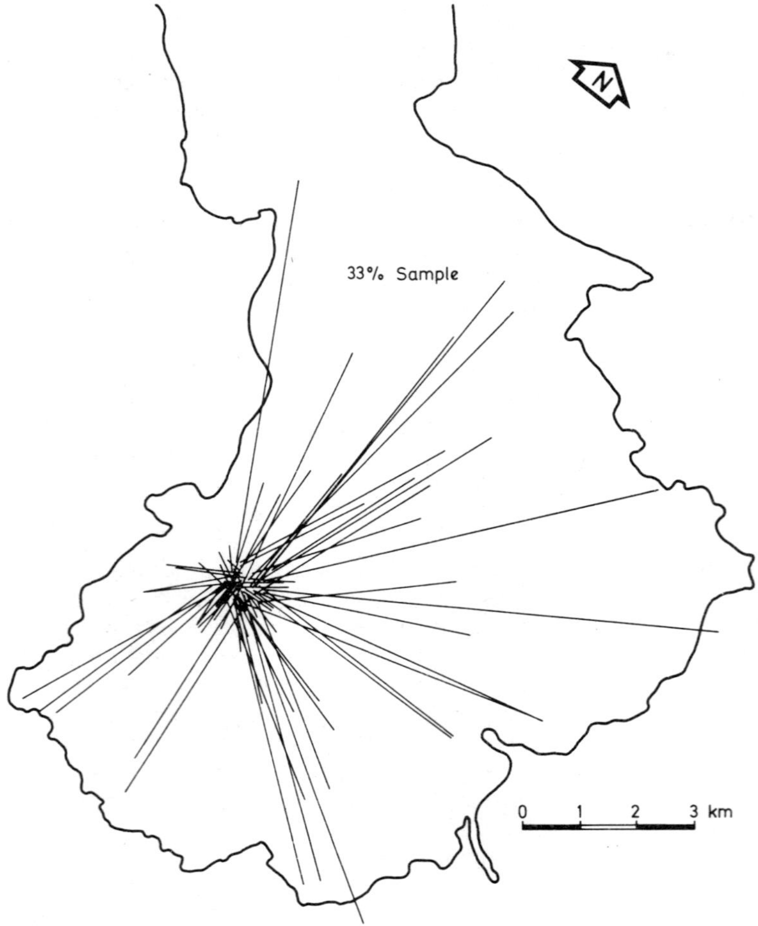

Figure 5.4 A sample of movement patterns to Bedminster

Figure 5.5 A sample of movement patterns to St. George

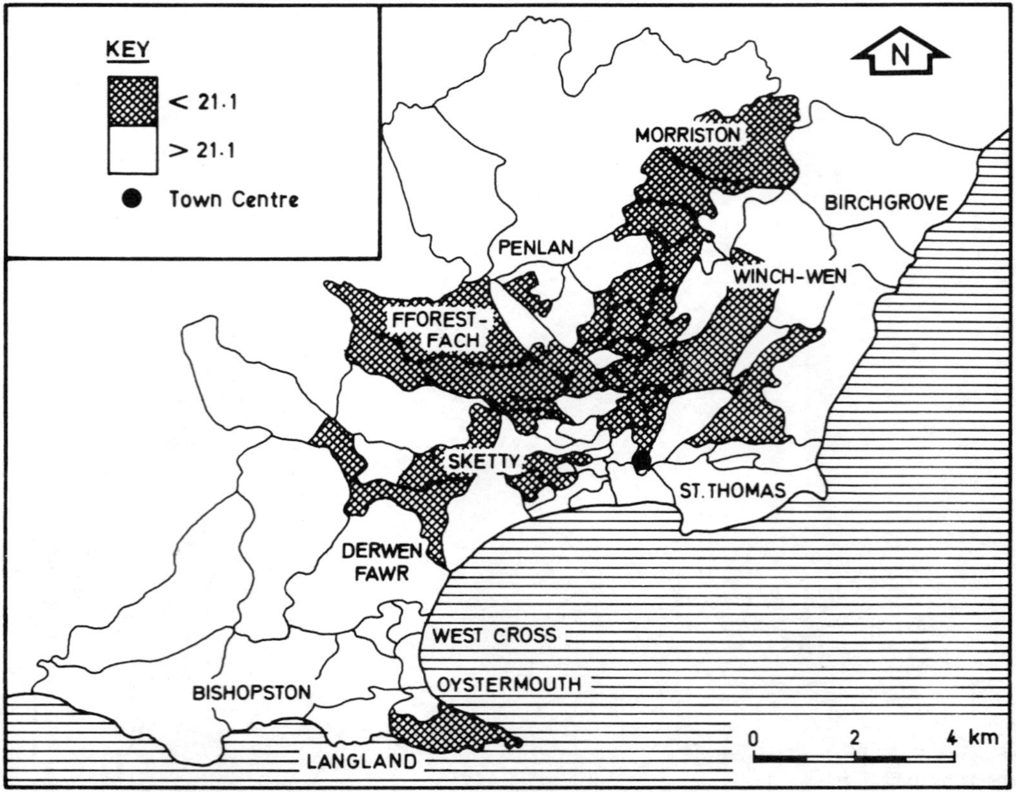

Figure 5.6 Rates of intra-urban mobility in Swansea (From Herbert, 1973)

this variable measures different types of movement in different areas of the city. In areas of contemporary house building this variable measures ***population growth***. But in areas, particularly in the inner city, with a constant population and a static housing stock the variable measures ***population turnover***. Unless one is aware of this property the intra-urban migration data should be treated with great caution.

5.3.4 Identifying movers and estimating population turnover

If one is interested in identifying recent movers, perhaps as a basis for questionnaire studies, and/or in estimating rates of population turnover, then the ***electoral register*** provides a convenient source of information. The electoral register is compiled annually from information obtained from a canvass held at the beginning of October each year. Local authorities are responsible for this canvass and send a registration form to each dwelling in their area. All households are obliged to give the name and address of all persons permanently living in the dwelling who are eligible to vote at the next election. The registers, which list the name and address of each person eligible to vote by streets, are first published in provisional form at the end of November. The registers are displayed in post offices and town halls so that people can check if their name and proper address has been included. In mid-December the provisional lists are removed, revisions and corrections are made and the final version is published in mid-February of each year. A copy of the register is deposited in the main library within the local authority.

The register obviously excludes those households who have moved between the period November to February. It also under-estimates the number of persons, and hence movers, in dwellings which exhibit multiple occupation since generally only one form is delivered to each dwelling and many households are thereby excluded. Despite these shortcomings the electoral register is the least biased sampling frame available to researchers in this country. Each register correctly identifies 95% of the eligible population at the time of publication, although this figure does fall to 85% just before the next register is published.

A comparison of these annual registers permits the identification of recent movers. *Table 5.4* shows by the use of fictitious names how the names and addresses are listed. From this example it is clear that Mr and Mrs Somebody have replaced Mr and Mrs Anybody at 16 Blank Street. This suggests that Mr and Mrs Somebody have moved into 16 Blank

TABLE 5.4 THE ELECTORAL REGISTER

1977 Register	1978 Register
Blank St.	Blank St.
16 A & B Anybody	16 E & F Somebody
14 C Jones	14 C Jones
12 D Smith	12 D Smith

Street between 1977 and 1978. By simply comparing the entries for the same addresses for the two most recent registers recent movers can thereby be identified. This procedure allows a sampling frame of recent movers to be constructed which can be used as a starting point for questionnaire surveys intending to identify areas of previous residence, tenure of previous residence and reasons for movement. Questionnaires should be phrased in such a manner which allow the differentiation between inter-regional, inter-urban and intra-urban movement.

A series of registers can also be used to calculate rates of population turnover. Short (1978b), for example, identified a group of households moving into eight areas of Bristol by comparing the 1963 and 1964 electoral registers. Persons appearing in the latter register but not in the former were identified as movers and the movers into each area were defined as a cohort. Successive registers up to 1975 were then analysed to discover the number in each cohort who were still remaining at the same address. For each area the percentage of the cohort remaining at the same address at each time period could therefore be calculated.

TABLE 5.5 POPULATION TURNOVER IN EIGHT AREAS OF BRISTOL

Area	Characteristics			Nos. in cohort	Half-life index	Terminal index
	Socio-economic status	Tenure*	Stage in life cycle			
1	Very high	O.O.	Child-rearing	587	3.8	24.1
2	High	P.R.	Pre-child	1245	1.7	6.2
3	Medium	O.O.	Child-rearing	107	6.9	41.8
4	Medium	O.O./ P.R.	Mixed	100	2.1	12.0
5	Very low	O.O./ P.R.	Pre-child	101	1.8	7.1
6	Low	O.O.	Post-child	100	6.0	35.2
7	Medium	O.O.	Child-launching	100	4.6	29.0
8	Low	L.A.	Child-rearing	664	4.5	25.2

*O.O. owner-occupation
P.R. private renting
L.A. local authority housing

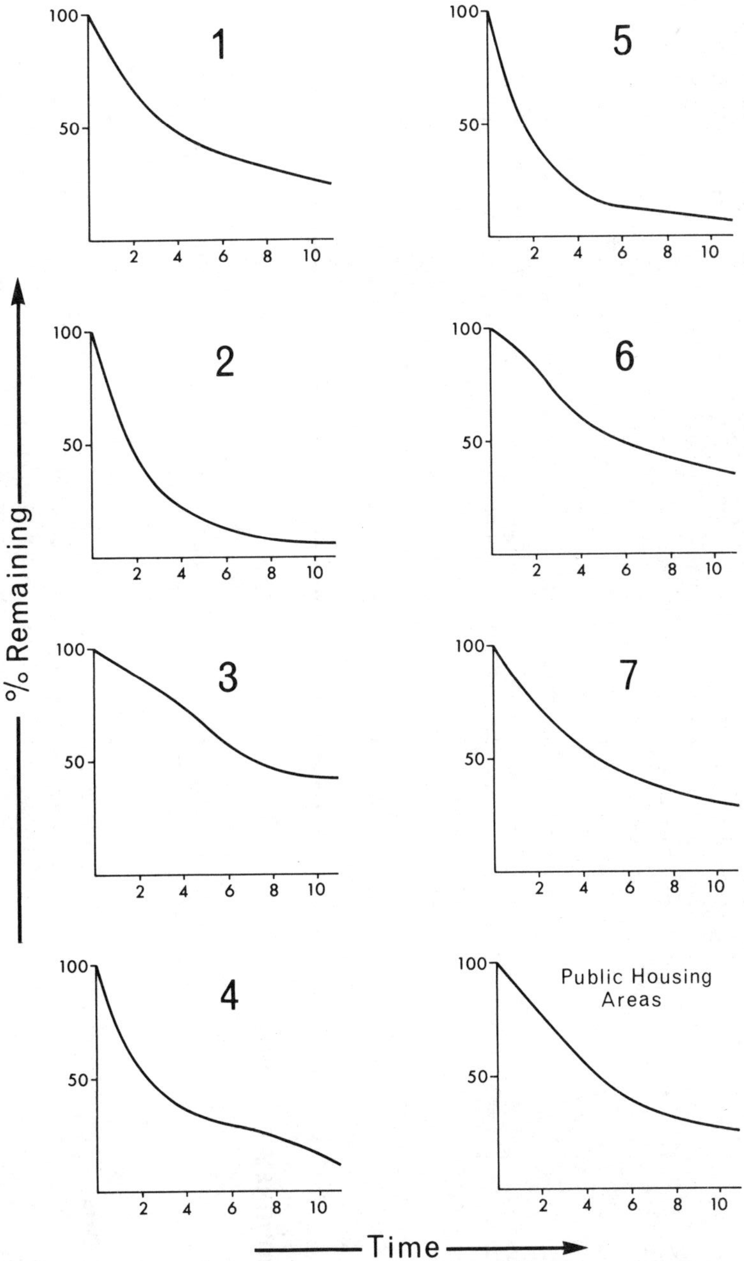

Figure 5.7 Population turnover in areas of Bristol

Table 5.5 lists the characteristics of each area, the numbers in each cohort and two indices. The half life index is the time taken for a particular cohort to lose half of its original number and the terminal index is the percentage of a cohort remaining at the same address at the end of the period 1963–1975. The percentage of each cohort remaining at the end of each year is shown for all the areas in ***Figure 5.7***. From this table and diagram it appears that high rates of turnover are associated with areas where private renting predominates and there is a majority of households at the early stages of the life cycle.

One advantage of this method of estimating population turnover is its degree of flexibility. The time scale of the analysis can be extended as far back as the available data. A series of time-staggered cohorts, for example, would allow the analysis of changes in turnover patterns. A wide variety of spatial scales can also be used, since the electoral register provides information for individual houses. The population turnover of particular blocks of flats or particular streets could be calculated. This flexibility enables us to use this method to answer very specific questions. For example in any block of flats is there any variation between the floors in terms of differences in population turnover? Alternatively, do roads with heavy traffic experience higher rates of turnover than streets with no busy traffic?

EFFECT OF EXTERNALITIES

5.4 JOURNEY TO WORK

The daily movement of people to and from work is an important facet of urban life. This movement uses up enormous amounts of time, fuel and in some cases also nervous energy. It is also part cause and part effect of the large concrete swathes of motorways which sweep through our major towns and cities.

5.4.1 Data

Information relating to various aspects of the journey-to-work phenomenon can be found in the ***Workplace and Transport Tables*** of the 1961, 1966 and 1971 Censuses. These tables provide information for each local authority on the economically active population – those living and working in the area and those resident in the area but working outside. A breakdown of the mode of journey to work by socio-economic status of residents is available for each ward and enumeration district of the city in the 1971 Census, Small

Area Statistics (*see Table 3.3*). Both these sources provide a fund of information on various aspects of journey to work. Since it is impossible to cover all the uses to which these data can be put, only two examples using the former source of data will be discussed in the following sections.

5.4.2 Suburbanisation

From the Workplace and Transport Tables of the Census it is possible to identify the number of persons working in the main urban centre in any particular region but residing in the surrounding local authority areas. Since these data are available for 1961, 1966 and 1971 they can, therefore, be used to show the increasing level of suburbanisation. For example from the relevant Workplace and Transport Table the percentage of the economically active population of surrounding local authority areas working in Bristol has been calculated for 1961 and 1971. The results are shown in ***Table 5.6***. The boundaries of these local authority areas are shown in ***Figures 5.8*** and ***5.9***. The table and associated figures show how the areas closest to Bristol were sending almost a half of their working population to work in Bristol even in 1961. Over the period 1961–1971 the more peripheral areas were increasing their proportion of working population going to work in Bristol. This increase reflects the increased ease of commuting. The figures also show how the sphere of influence of Bristol, defined in this case in terms of levels of commuting, increased over the 10-year period.

TABLE 5.6 JOURNEY-TO-WORK STATISTICS, 1961–1971

Area	**Percentage of economically active population working in Bristol**	
	1961	**1971**
Kingswood	47.2	48.6
Warmley	41.2	41.9
Keynsham	26.2	41.3
Mangotsfield	49.6	48.8
Portishead	21.9	30.4
Long Ashton	38.8	44.6
Sodbury	26.6	27.5
Thornbury	19.3	23.7
Clevedon	13.8	19.8
Clutton	13.7	20.5
Weston-Super-Mare	7.2	15.5
Axbridge	6.1	12.9
MEAN	26.4	32.3

This index of commuting — the percentage of the economically active population living in surrounding areas but working in the main city — has been used to define city regions. In the United States the Standard Metropolitan Statistical Areas (SMSA) are defined by including all local areas which have more than 15% of the economically active population working in the major urban centre. These SMSA's are widely used for transport and planning schemes and as basic data collection units. Hall ***et al***. (1973) have used the same threshold value to define Standard Metropolitan Labour Areas (SMLA) in England and Wales.

5.4.3 The differential nature of suburbanisation

The journey-to-work data which we have just used is also available for the different socio-economic status categories. In the 1961, 1966 and 1971 Census, Workplace and Transport

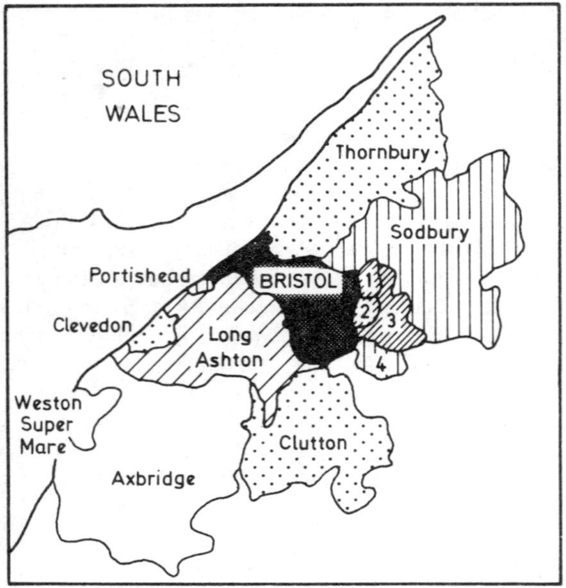

1 Mangotsfield
2 Kingswood
3 Warmley
4 Keynsham

KEY TO % OF ECONOMICALLY ACTIVE POPULATION WORKING IN BRISTOL

+40 %
+30 - 40 %
+20 - 30 %
+10 - 20 %

Figure 5.8 Journey-to-work patterns, 1961

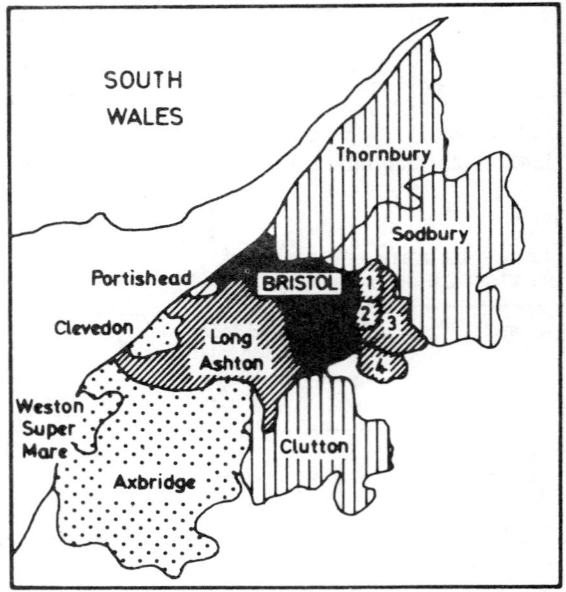

1 Mangotsfield
2 Kingswood
3 Warmley
4 Keynsham

KEY TO % OF ECONOMICALLY ACTIVE POPULATION WORKING IN BRISTOL

+40 %
+30 - 40 %
+20 - 30 %
+10 - 20 %

Figure 5.9 Journey-to-work patterns, 1971

Tables the numbers residing and working in a city (WIRI), the numbers residing outside but working inside the city (WIRO) and the numbers working outside but residing inside the city (WORI) are available for each socio-economic status category. From these statistics Thomas (Hall ***et al.***, 1973, Vol. 2, Chapter 9) has devised an index of home–workplace separation (IhwS) where,

$$\text{IhwS} = \frac{\text{WIRO} - \text{WORI}}{\text{WIRI} + \text{WIRO}} \times 100$$

A high value on this index reflects a large separation between home and workplace and the socio-economic group with the highest value has the greatest separation between home and workplace, i.e. it is the most suburbanised socio-economic status category.

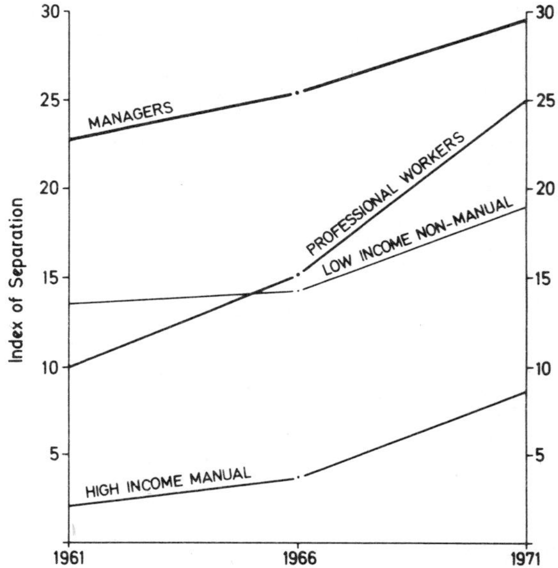

Figure 5.10 Home–workplace separation by socio-economic status

Taking Bristol as the main city and the areas mentioned in ***Table 5.6*** and ***Figure 5.8***, the IhwS index has been calculated for five socio-economic categories for the periods 1961, 1966 and 1971. The results are portrayed in ***Figure 5.10***. Although all the socio-economic status categories have experienced increased separation between home and workplace between 1961 and 1971 the results show that this separation is highly selective. Those in non-manual occupations are more 'suburbanised' than those in manual occupations. A number of reasons can be put forward to explain this pattern. In the suburban areas surrounding major cities the bulk of the housing is owner-occupied. The Building Societies, whose mortgage allocations allow households to become owner-occupiers, tend to grant more mortgages to those in non-manual occupations. Since non-manual workers, especially the higher income managers and professional workers, have higher incomes than manual workers they can more easily afford the cost of car ownership which is essential for long distance commuting. The differential nature of suburbanisation shown in ***Figure 5.10*** can, therefore, be seen as a reflection of the unequal distribution of income and of the lending policies used by the Building Societies in allocating mortgages.

CHAPTER 6 THE INDUSTRIAL AND COMMERCIAL STRUCTURE OF THE CITY

6.1 INTRODUCTION

In this chapter we will focus on a range of data sources and a number of concepts which enable us to describe the industrial and commercial structure of the city. This topic is worthy of attention since it is cities wherein the bulk of manufacturing, the exchange of commercial information and the provision of retail goods and other services takes place. Such functions provide the ***raison d'etre*** of cities.

6.2 INDUSTRIAL LAND USE

6.2.1 General considerations

Before discussing the location of different types of industry it is important to consider the location of industry with respect to other types of land users (such as, commercial land use, retail land use and open spaces) within the city. Such a comparison is possible from an analysis of the plans available from the local authority planning departments. For most local authorities there are one, and perhaps two, major sets of plans referring to the post-war period. Firstly, there are the ***development plans*** which were required by the 1947 Town and Country Planning Act. This Act required local authorities to indicate, by a series of maps and written statements, a plan of proposed land uses in the city for the next 20 years, subject to 5-year revisions. In their development plans most local authorities produced land use maps which indicated the broad, and sometimes detailed, patterns of current and proposed land use. Secondly, the 1968 Town and Country Planning Act has required all local authorities in England and Wales to produce a more up-to-date ***development plan*** consisting of a ***structure plan***, which is a written statement dealing with the broad sweep of planning policies, and ***local plans*** which consist of written statements, maps on an ordnance survey base and other diagrams which provide more precise details of the proposals mentioned in the structure plan.

The maps and diagrams associated with these various development plans are open to public inspection at the local planning department and are regularly shown in planning exhibitions. A perusal of these maps and diagrams is strongly recommended. They provide an important source of data as well as a guide to present and future planning proposals. As an exercise, collect all the planning proposal documents related to your town or city. Do you agree with them? Can you improve them? If you disagree with them make representations to your local planning department or local authority council.

The uses to which these local authority development plans can be put are manifold. However, in the present context we are primarily concerned with them as a source of data referring to the location of different types of land use. The land use maps associated with the development plans provide plentiful information on the distribution of different land uses. From an analysis of such maps isolate the distribution of industrial land use within your town or city. The results of such an examination can be compared to the patterns discussed in the next section.

6.2.2 Industrial land use in British cities

In most major British cities industrial land use will generally have the following distribution:

(a) An inner ring of industrial land use encircling the commercial and retail core of the city.
(b) A 'ribbon' form of development from the central city outwards along certain important transport routes.
(c) A suburban or exurban ring of new factory developments, often on greenfield sites.

The sites associated with (c) are often due to the relatively recent construction of industrial estates by either the central government or by the local authorities. Since 1947 and, until very recently, the main thrust of land use planning in Britain has been directed towards segregating industry from other, especially residential, land users. To achieve this aim the planning authorities have endeavoured to remove manufacturing firms in the inner city to the greenfield sites of new trading estates. However, such an aim is increasingly being questioned. In the first place it removes job opportunities from the central city and further disadvantages those workers in the inner city who do not own a car. In the second place such an aim was clearly desirable when industry was noisy, dirty and noxious, but the adoption of more modern industrial processes has lessened the incongruous nature of industry being located within or beside residential areas. In the third place the separation between home and workplace, which is one consequence of industry being located on greenfield sites, increases the cost and time of journeys to work and puts further strain on the transport networks.

Do the planning proposals related to industrial land use in your city attempt to segregate industrial land use from residential land use?

6.3 THE INTRA-URBAN LOCATION OF INDUSTRY

6.3.1 Data sources

There are a number of data sources which allow us to identify the spatial distribution of different types of industry. One of the best is ***Kelly's Directories***. In 1844 F.F. Kelly left his job at the Post Office to set up his own directory. From 1845 to 1946 ***Kelly's Directories*** were published for most of the counties in this country and for most of the major provincial towns and cities. From 1946 Kelly's ceased publication of the county directories but they still issued directories for the main towns and cities in Britain on an annual basis up until 1976, although in some cities the period between successive directories can exceed 1 year. The most important part of the directories is the list of the addresses of firms and shops in the area. These firms are listed alphabetically under the name of the company or firm (***Figure 6.1a***) and by the nature of the manufacturing, retail or service function (***Figure 6.1b***). The successive ***Kelly's Directories*** for a particular town or city can be found in the reference section of the local public library.

A comparable data source is the ***Yellow Pages*** of the Telephone Directory. However, because some establishments prefer not to enter their name in the ***Yellow Pages*** and, since this source does provide the longitudinal data available in Kelly's, this suggests that both ***Kelly's Directories*** and the ***Yellow Pages*** need to be consulted.

The official data sources on industrial location are more reliable, but unfortunately more inaccessible. The records of the local ***Factory Inspectorate*** provide information on each industrial plant or firm including the name and address, the date of opening and the number of workers employed. This last piece of important information is not available

Broadmore & Cook (Building Contractors) Ltd. Stanley st
Telephone No Reading 57612

Breakout at Jackson, tailors, 46 Butts centre

Breeze Derek R. BA. solicitor (firm. Sarjeant. Brown & Co), 150 Friar st RG1 1HF. Tel No 53425

Brenda, ladies' hairdrssr, 277a Basingstoke rd

Brennan & Sons, clothiers, 7 London st

Brent Bros. Ltd. outfitters, 4 Oxford rd

Brentford Nylons, soft furnishers, 30 Station hill

Brereton J. H. grocer. 6/8 Park la. Tilehurst

Brewery Tavern, 19 Caversham rd

Brian J. Woods Ltd. photographers. 310/312 Oxford rd

Brian Rawling, do-it-yourself supplies, 14 Norcot rd, Tilehurst. Tel Reading 24444

Bridge Boat Ltd. Frys Island, Thames side

Briggs W. & Co Ltd. motor tyre distributors. 646 Oxford rd

Bright's, estate agts, 116 Butts centre

Bristol & West Building Society, 173/174 Friar st

Bristow R. E. Ltd. precision engnrs. Darwin close

Bristow Herbt. grocer. 33 Wolseley st

Britannia Hotel, 50 Caversham rd

Britannia Tap P.H. 48 Caversham rd

Britannic Assurance Co. Ltd. 55 Minster st

British Airways, travel agts, 2 St Mary's butts

British Council (The), 44 Addington rd

British Domestic Appliances (Service Division). 13 Cremyll rd

British Estate Services Ltd, public works contrctrs, 132 Bath rd

British Estate Services (Properties) Ltd, concrete products, The Meadway, Tilehurst

British Fermentation Products Ltd. 55 Loverock rd

British Home Stores Ltd. departmental stores. 116/119 Broad st. & 23 Minster st

British Insulated Callender's Cables Ltd, cable mfrs, Unit, 20/22 Bennet rd

British Legion (Earley & Reading branch), 31 Auckland rd

British Legion (Royal) (The) (Royal County of Berkshire) (R. E. Shrive BEM, county sec), Watlington ho, 44 Watlington st

British Nursing Association, 18 Cross st

British Oxygen Company (sales centre), Rose Kiln la

British Olivetti Ltd, typewriter mfrs, 6/8 Arkwright rd

British Rail Travel Centre, bookings for most entertainments, holidays & travel, 40 Station hill. Tel 55977

British Railways Hotels Executive, refreshment rms. Western Region Railway station. Station approach

British Railways (Western Region) Reading Station; London Division Headquarters, Divisional Civil Engineer, Chief Accountant, Chief Signal & Telecommunications Engineer, Western Tower, Station hill; Reading West Station, Oxford rd; Tilehurst Station, Oxford rd; Freight Depot, Vastern rd; Coal Yard, Willow st & Diesel Depot, Cow la

British Railways (Western Region) Staff Association Institute & Club, Station hill

British Red Cross Society (County Branch Headquarters) (Group Capt W. J. Swift CVO, branch director), 48 London rd

British Red Cross Society (Reading Division) (Mrs P. H. Pettitt, divisional sec), 48 London rd

British School of Motoring Ltd. 12 & 14 Blagrave st

British Shoe Corporation, boot mkrs. 13b Union st

BROADBEAR BROS. LTD. sheet metal workers, 92 Audley st Tel Reading 53970

Broadway & Malyan, chartrd archtcs. 101 London rd

Brock H. L. grocer. 36 Collis st

Brook Street Bureau of Mayfair Ltd. employment agencies, 60 Broad st

Brook Street Technical & Industrial Services Ltd, employment agencies, 60 Broad st

Brooklands Autos (Reading) Ltd. motor engnrs, James Garage, 330 Tilehurst rd

Brooks R. (Builders) Ltd, building contrctrs, 33/35 Greyfriars rd

Brown Brothers Ltd. electrcl. wholesalers. 71/73 Caversham rd

Brown Peter, men's wear, 15/17 Butts centre, Oxford rd

Brown C. S. fruitr. 137 Oxford rd

Brown F. R. shopkpr. 1 Rupert st

Brown J. motor cars for hire, 24 Edgehill st

Brown R. (Chemists) Reading Ltd. chemists, 266 & 268 Oxford rd

Brown R. W. baker, 89 London st

Brown W. & Sons (Reading) Ltd. coal mers. 651 Oxford rd

Brown Wilfred G. newsagt, 5 Castle st

Brownlee George & Partners, chartered quantity surveyors, 32 Queen's rd RG1 4AU. Tel 583236

B.R.S. PARCELS LTD,
carriers, Cardiff road RG1 8HW. Tel Reading 53404/5

Brunning Advertising & Marketing (Reading) Ltd. advtsng. agts. 9 King's rd

Brunsdon N. A. newsagt. 23 London rd

Bruno M. hairdrssr, 95 London st

Bruno M. hairdrssr, 89 Southampton st

Brunswick Arms P.H. 49/53 Brunswick st

Brunswick International (UK) Ltd, sports equipment wholesalers, Macgregor ho, Paddock rd, C

Brunton D. turf accntnt, 11 The Triangle, T

Brunton Dick Ltd, turf accntnts, 209 Caversham rd

Brunton R. turf accntnt. 123 Queen's rd; 328 & 737 Oxford rd & 233 London rd

Brush Beryllium Ltd. 20 Boulton rd

Brush Engineering Co Ltd, genl engnrs, 37/40 Minster st

Brush Mallory Ltd, suppliers of beryllium products, 20 Boulton rd

Bryan J. Ltd. ladies' outfitters. 10 Yield Hall pl

Buchanan Wltr. J. shopkpr. 22 Chester st

BUCKLAND & SONS, surveyors, valuers, auctioneers & estate agents, 7 Blagrave st (Telephone No 51370): & at Windsor, Slough & Gerrards Cross

Budgen & Co. Ltd. grocers. 168 Whitley Wood rd. & 80 Christchurch rd

Budgen W. E. J. & Partners. civil engineering consultants, 47 Crown st

Bugle P.H. 144 Friar st

Bull Hotel, 23 Broad st

Bull Howard Ltd, advtsng agts, 5 Cross st

Bullion Bros Ltd, painters. 107 Caversham rd

Bullion & Colebrooke, french polishers, 239 Southcote la. Southcote

Bundy Plastic Products Ltd, glass fibre products, 21 Milford rd. Tel 583196

Bunting J. S. MB. DMRT. consultant radiotherapist. 1a Erleigh rd

Bunting R. greengro 52 Prospect st. C

Burberry Reading Ltd, rainwear manufacturers, Mill la

BURGESS & CO (ENGINEERS) LTD,
heating & ventilating engineers, Easthampstead rd, Bracknell Tel 0344 21341, Telex 847520

Figure 6.1a and b Extracts from *Kelly's Directory* of Reading, 1975

Coin Dealers

Coin & Stamp Centre, Reading (valuations given), 31 Castle st. Tel Reading 55593
Collectors Bazaar (Reading), 631 Oxford rd. Tel 581960
Numismery (The), 189b Oxford rd. Tel 582145
Thames Valley Coins, 32 Queen's rd. Tel 51646

Cold Storage

Reading Ice & Cold Storage Co (The), 32 Lynmouth rd. Tel 52024

Colour Manufrs

Reading Green & Marvell Ltd, 34 Kennet side & 117 Loverock rd

Commercial Artists

Arnold Poster Advertising, Prospect mews, Prospect st. Tel 54585
Bower John Studio Ltd, 23 Loverock rd, Battle Farm Trading Estate RG3 1DZ. Telephone, Reading 53819
Oryx Design, 55a Baker st (Tel 581606) & 177a Wokingham rd. Tel 666961
Searle Studio, 50 Mount Pleasant. Tel 83578

Commercial Photographers

Brian J. Woods Ltd (industrial, commercial & injury photography a speciality, also colour processors to the trade), 310/312 Oxford rd RG3 1AD. Tel Reading 583888; STD Code 0734; (after hours) Reading 690335
Mills W. M. (wedding photographs a speciality), 79 Christchurch rd. Tel Reading 81511

PHOTOGRAPHIC SERVICES (BERKSHIRE) LTD
(distributors of 3M printing products), 11 Boulton road RG2 0NE. Phone, Reading 81686/7

Commercial Stationers

Reading Office Supplies, 14 Church st, Caversham RG4 8AU. Tel 478022
Taylor Rt & Co, 83 King's rd. Tel 53728

Commercial Vehicle Dealers

GREAT WESTERN MOTORS (READING) LTD,
(Bedford main dlrs), 57 Vastern rd. Tel 55501 (9 lines). See our advertisement in Kelly's Shop Window at the beginning of the Tradefinder

Commissioners for Oaths

See Solicitors

Computer Consultants

Computer Analysts & Programmers (Reading) Ltd, New Oxford ho, 1 Station rd. Tel 580264
Tylin Management Systems Ltd, Fernbrook The Square, Pangbourne. Tel 3838

Computer Manufrs

Digital Equipment Company Ltd, 3/4 Arkwright rd (Tel 583555) & Butts Centre, West Mall
International Computers Ltd, 30/31 Friar st & Cardiff rd. Tel 581258

Computer Peripheral Equipment Manufrs

P.E.C. Ltd (Pertec), 10 Portman rd, Battle Farm Estate
Tally Ltd, 7/9 Cremyll rd. Tel 580141

Concrete Mixer Manufrs

Ritemixer Ltd, Darwin cl. Tel 860124

Concrete Products Manufrs.

British Estate Services (Properties) Ltd, The Meadway, Tilehurst. Tel 54540
READING CONCRETE PRODUCTS LTD
(precast concrete manufacturers), Bridge works, Berkeley avenue. Telephone, Reading 50619

Concrete Suppliers—Ready Mixed

Era Concrete Pumping Services Ltd, 5 The Fairmile, Henley-on-Thames, Oxon. Tel Henley-on-Thames 3801. See advert
Folley's Ready Mixed Ltd, Spade Oak Plant, Marlow. Bourne End 24741/3. See advert
Premix Concrete Co Ltd, Old Bath rd, Twyford (Tel Twyford 340389) & Burghfield rd, Theale. Tel Reading 302535
Ready Mixed Concrete (Thames Valley) Ltd, 472 Basingstoke rd RG2 0QJ. Tel 85522

Confectioners—Retail

Bacon Miss W. 33 Wilson rd
Bailey Mrs C. 286 King's rd
Batsford D. E. & K. J. 25 Church rd, Caversham. Tel 473187
Bolton D. 59 Southampton st. Tel 55300
Bowen W. R. 101 Wokingham rd. Tel 63879
Brunker P. J. & J. R. 5 Royal av, Calcot. Tel 28778
Candy Box, 46 Caversham rd. Tel 51531
Cannings of Reading Ltd, 33/35 Armour rd, Tilehurst (Tel 25373) & (head office) 252 Shinfield rd. Tel 84051
Chocolate Box, 65 Whitley st
Cooper Mrs O. F. H. 263 Oxford rd. Tel 53747
Deverall H. 203 Hall's rd, Tilehurst. Tel 28834
Ford Miss G. 253 London rd
Gale H. & E. 87 Erleigh rd. Tel 65474
Georgiou A. 83 Vastern rd. Tel 50724
Gordons, 161 Friar st
Green W. V. & C. 92 School rd, Tilehurst. Tel 27295
Griffin F. E. & B. A. 441 Basingstoke rd. Tel 82016
Hamilton Miss Violet, 33 Kensington rd. Tel 52646
Hawkey J. H. & P. M. 645 Oxford rd
Hurley Mrs D. C. 63 Gosbrook rd, Caversham
Jane B. W. 188 King's rd
Johnson A. J. P. 32 Erleigh rd. Tel 62589
Johnson D. & A. (also tobacconists & greeting cards), 197 London rd. Tel 65606
Koba J. H. 15 Prospect st, Caversham. Tel 471479
Lavells Ltd, 21 West st
Lewis J. H. 39a Berkeley av
Martin R. W. 198 Oxford rd. Tel 53175
Matin R. A. & R. A. E. 133 Wokingham rd. Tel 62560
Maynards Ltd (Naylors), 6 The Arcade. Tel 54795

in either ***Kelly's Directory*** or the ***Yellow Pages***. However, these official data are not open to public inspection and prior permission to consult the data has to be obtained from the local ***Factory Inspectorate.***

6.3.2 Location by type of industry

Industry is not randomly located throughout urban areas. Different types of industries with varying locational requirements will be attracted to different parts of the city.

A generalised picture of intra-urban distribution of industry is summarised in ***Figure 6.2***. The 'ideal' city pictured in ***Figure 6.2*** has four different types of industry classified with reference to location from the city centre. Type A industries, located close to the

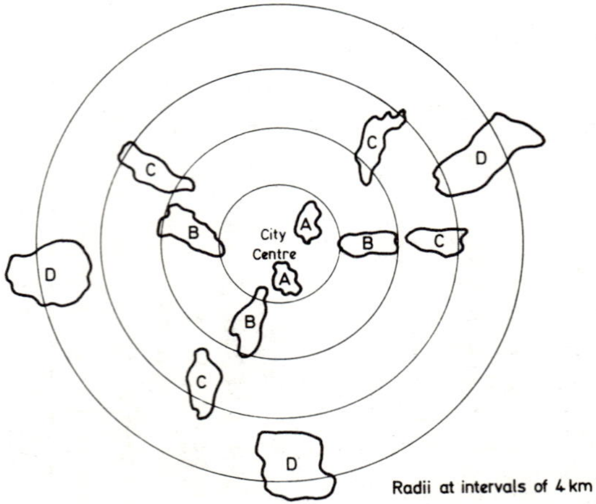

Types of industry

A Printing, newspapers, clothing, office machinery
B Packaging, distribution
C Food manufacturers, engineering, electronics
D Vehicles, heavy engineering, oil refining

Figure 6.2 A model of intra-urban industrial location (After Hamilton, 1967)

city centre, consist of three main types. Firstly, there are those industries such as printing and newspapers which require the accessibility to skilled labour and the maximum accessibility for the distribution of their products provided by a central city location. Secondly, there are those industries, such as office machine manufacturing, which are attracted to the market provided by the nearby location of offices in the commercial core. Finally, there are those industries — generally small scale concerns — which are located in the inner city simply through inertia; initially attracted to the central city for a variety of reasons which are now no longer operative, they remain in the central city because of their inability to face the high costs of relocating to more optimum locations. It is these latter types of industry which are experiencing greatest difficulty by land use planning schemes on the one hand and general economic difficulties in a competitive market on the other.

Type B industries, located around the fringes of the central core, are those industries which require the accessibility provided by a central city location but are unable to afford a prime city centre site because of the prohibitive land costs. Such industries include packaging and wholesale distributors.

Type C industries located further out from the city centre include the larger, more modern industries such as light engineering which seek cheaper land and easy access to labour. Such industries often locate on radial transport routes. Finally, there are type D industries located at the periphery of the city. These industries are those modern, large scale industrial concerns, such as oil refining and vehicle manufacturing, which require large supplies of cheap land.

The picture presented by ***Figure 6.2*** is only a general outline which does not take into account the industrial structure of cities in different parts of the country, the differences between towns of varying sizes or such complications as port industries. However, the schema does highlight the fact that industries which are older and which require accessibility will generally be found in the central city, while the more modern and larger scale industries will be found further out from the central city. Using ***Kelly's Directories*** plot the present location of different types of industries in your town or city. Does the pattern correspond to the situation outlined in ***Figure 6.2***? If not, why? Relate the distribution of the industries you consider to their varying locational requirements.

Certain types of industry will show a tendency to cluster together. For example Wise (1949) has shown that the gun and jewellery manufacturers in Birmingham were concentrated in one particular quarter of the city. In a subsequent section of this chapter we will highlight a number of techniques which allow the degree of clustering between retail establishments to be measured. These techniques, used on the data available in ***Kelly's Directories***, can be similarly applied to the clustering of industrial establishments.

6.3.3 Patterns of change

Although the high costs of relocation prohibit the continual movement of industrial establishments, such movement does occur. Spatial patterns of industrial location do not remain constant and the spatial distribution of industry changes, albeit slowly, in response to a variety of factors. In this section we will be more concerned with the patterns of movement rather than the reasons for such movement, since there are a number of textbooks which adequately discuss this latter topic.

General remarks

Before focusing on three aspects of industrial movement through the use of ***Kelly's Directories***, it is necessary to introduce some terms. Imagine that for one particular city we have obtained copies of ***Kelly's Directories*** referring to two different time periods; time t and time $t + 1$. For any particular firm mentioned in these directories we can identify four different categories. If we find that the same firm is located at the same address at time t and at time $t + 1$, then this firm is classified as a ***stayer***. If we find that a firm has moved its location, the entry in the directory gives a different address in time $t + 1$ from that of time t, and this firm is then referred to as a ***mover***. Alternatively, if we find that a firm is recorded in time t but is not recorded in time $t + 1$, then it is classified as a ***death***. Conversely, if we find that a firm is not recorded in time t but is recorded in time $t + 1$ this is classified as a ***birth***. These four categories are represented in diagrammatic form in ***Figure 6.3***. With these simple categories we can, by using ***Kelly's Directories*** and ***Yellow Pages***, analyse three aspects of industrial movement.

Movement by sector of industry

Since ***Kelly's Directories*** list firms involved in the same branch of industry under the same

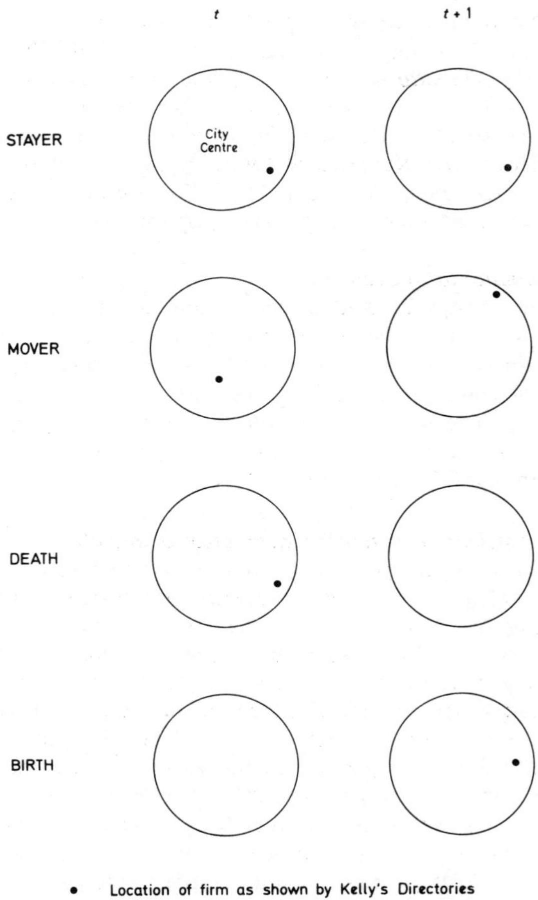

Figure 6.3 Four categories of firm relocation

TABLE 6.1 MOVEMENT OF FIRMS IN SHEFFIELD (From Bull and Bull, 1978)

Industry	Movers as percentage of stayers
Cutlery	42.1
Printing	46.1
Cabinet-making	14.2

heading then successive directories allow us to note the changing distribution of different sectors of industry. For example a study of the Sheffield cutlery industry which used ***Kelly's Directories*** for that city found that, between 1951 and 1973, there were 18 births of cutlery firms, 28 deaths and 6 transfers; only two firms remained in business throughout this time period (Bull and Bull, 1978). The same article also provides information, again derived from ***Kelly's Directories***, on the amount of movement by firms in different industries over the period 1948–1973. The results shown in ***Table 6.1***, highlight the fact that firms in some sectors of industry are more prone to movement than others.

Direction of movement

Since ***Kelly's Directories*** give the present and previous addresses for all moving firms within one particular city, then these data can be used to plot the direction of movement of firms. Given the previous remarks we could suggest that the predominant form of movement will be outwards from the city centre as redevelopment schemes have caused many firms to relocate further out from the city centre in the more peripheral trading estates. Test this proposition by mapping the direction of movement of firms moving within your town or city.

Industrial movement between areas of the city

Over any given time period we can use the information in ***Kelly's Directories*** to ascertain the pattern of movement between different areas of the city and the consequent changing economic fortune of different parts of the city. This can be achieved by identifying the number of births, deaths and movers in different parts of the city. For example Lever (1972) has used the Post Office postal directories to identify such firms in four zones of Glasgow; the results are shown in ***Table 6.2***. The inner city zones 1 and 2 experienced a net loss of firms between 1959 and 1969 as deaths exceeded births, but the outer city zones 3 and 4 experienced a net gain of firms as the number of births greatly exceeded the number of deaths. The greatest net gain of firms was recorded in zone 4. Overall, these results suggest that there is an outward movement – a suburbanisation – of industry in Glasgow. If these trends continue it can be suggested that eventually there will be very little industry in the inner city with a consequent reduction of job opportunities for inner city residents without cars.

TABLE 6.2 BIRTHS AND DEATHS OF FIRMS IN GLASGOW (From Lever, 1972)

Zone	Distance from city centre	A Births*	B Deaths†	A–B Net balance
1	0–3	46	50	–4
2	3–4	43	25	–18
3	4–6	32	8	+24
4	6–10	65	10	+55

*Includes firms who moved in from other zones
†Includes firms who moved out to other zones

Using ***Kelly's Directories*** for 1959 and 1969 and a four-zone framework similar to Lever's, examine the births and deaths of firms in your town or city. Do the patterns you discover correspond with Lever's? You may have to adopt a zone framework more suitable for the size of town or city which you are considering.

6.4 THE RETAIL AND COMMERCIAL STRUCTURE OF CITIES

6.4.1 Data sources

In this section we will focus attention on the retail and commercial structure of cities. Such an analysis can be based on the data sources which we have just been considering. For example, the development plans available from the local authority provide information on the present and proposed patterns of retail and commercial land use; ***Kelly's Directories*** and ***Yellow Pages*** also provide information on retail and commercial firms as well as manufacturing firms. However, there are a number of other data sources which can also be used.

Rateable value records

The gross rateable value of each and every property in any particular town or city can be obtained from the rates office of the local authority where the records are freely available for public inspection. The records are published by wards of the town or city. ***Table 6.3*** provides selections from the rateable value records taken from two streets in Bristol.

TABLE 6.3 EXTRACTS FROM RATEABLE VALUE RECORDS

Description	Address	Gross value (£)	Rateable value (£)
Congregational Church	36 Stapleton Rd	148	97
Shops and premises	40 Stapleton Rd	325	245
House	42 Stapleton Rd	150	98
Flat 1st floor	44 Stapleton Rd	92	60
Maisonette flat	44 Stapleton Rd	164	117
Two offices, ground floor rear	10 St Augustine's Parade	265	194
Offices and premises, 1st floor rear	10 St Augustine's Parade	330	250
Offices and premises, 1st floor front	10 St Augustine's Parade	290	215
Offices, 2nd floor	10 St Augustine's Parade	240	173
Offices, 3rd floor	10 St Augustine's Parade	130	82
Shops and premises	10 St Augustine's Parade	3700	3055

We can see from these tables that the records give the address of the property, whether it is used for non-commercial or commercial use, the gross rateable value and the rateable value. The gross rateable value is assessed by a valuation officer employed by the local authority. The assessment considers the physical condition of the property (including the amount of space), the frontage and accessibility and the potential value of the site on which the property is located. The gross rateable value is considered by the assessment officer to be a 'fair' annual rent which could be charged for the property; it is thus an indirect measure of land value. It must be borne in mind that the gross rateable value reflects the views of the assessment officer and is not a market price. The rateable value figure on the extreme right in ***Table 6.3***, is the figure used to calculate the rates which the owners or occupants of the property must pay to the local authority. The rateable value is less than the gross rateable value and this difference is the likely annual cost of the maintenance of the property. The rates charged by local authorities vary throughout the country, but if the rate is fixed at 80 pence in the £ then the owners or occupiers of 42 Stapleton Road (***see Table 6.3***) would pay (0.80 X 98=) £78.40 each year in rates. In theory the gross rateable value of properties is supposed to be assessed every 5 years but, in practice, this rarely occurs. The last assessment took place in 1973 and the next assessment is scheduled for 1982.

Shopping centre plans

Since 1967 a large number of maps of shopping centres in British towns have been published by the map publishers Charles E. Goad Ltd.* This company has produced maps of the shopping areas of all major towns and cities in Britain with a population of over 50 000. Plans have also been published of a number of suburban shopping areas in the larger towns and cities. In Bristol, for example, ten different shopping areas have been mapped.

The coverage of these Goad maps can be seen from the portion of the Goad shopping plan of central Bristol shown in ***Figure 6.4***. The maps use a scale of 88 ft to the inch, and indicate street numbers, the area and frontage of properties, the name of the occupant and the type of retail activity. Subsidiary information provided by these maps include the total population of the town or city, the distance from London and the early closing day. The maps cover the area of continuous retail activity but, as Rowley and Shepherd (1976) have pointed out, this does not exclude the possibility of arbitrary cut-offs in the areal extent of map coverage. One further drawback is that the maps only include information on the occupants of the ground floor of properties.

Since these maps are published every few years, a comparison of the maps of the same shopping area for different time periods can highlight the turnover and movement of different retail establishments. More recent movement of retail establishments can also be highlighted by using the most up to date shopping map as a basis on which to conduct a variety of fieldwork exercises.

The various data sources which we have considered can be used to describe various facets of the retail and commercial structure of cities. In the following sections we will

*Charles E. Goad Ltd., 18a Salisbury Square, Old Hatfield, Hertfordshire AL9 5BE

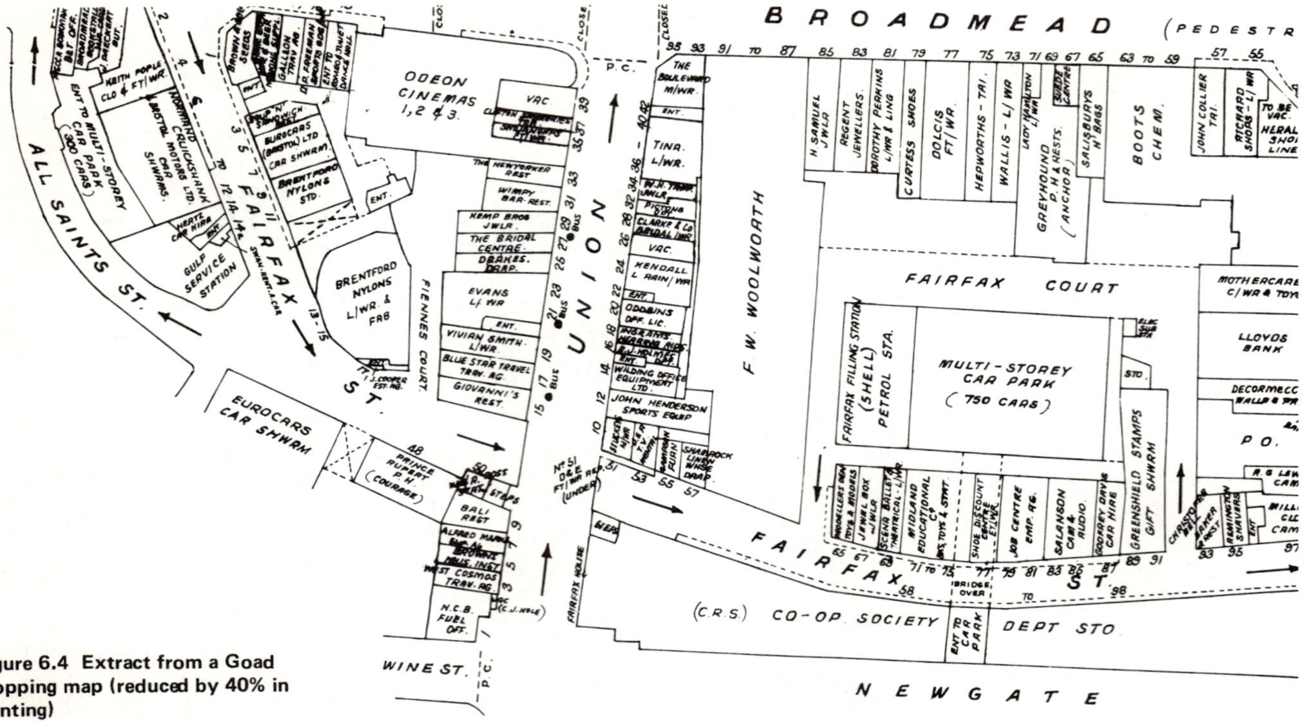

Figure 6.4 Extract from a Goad shopping map (reduced by 40% in printing)

consider three main facets: (1) land values and accessibility, (2) the central business district and (3) the spatial distribution of various retail and commercial establishments.

6.4.2 Land values and accessibility

Many retail and commercial firms need to be in accessible locations in order to attract customers. Since there will be many such firms in any one particular city this causes competition for accessible sites. Through the process of competitive bidding for sites we would expect land values and land rent — the price paid to a landowner for the use of a site — to be greatest in areas of maximum accessibility. We would, therefore, expect those firms which can afford to pay the higher rents to be located in areas of maximum accessibility. We can express these relationships by way of a simple example. Consider three types of land users: retail shops, commercial firms and households seeking dwellings. The retail shops require accessible sites in order to attract customers. The more accessible their location, the more customers are attracted and the greater the profit; conversely, the less accessible the location, the fewer the customers and the smaller the profit. Because rents are paid out of profits, retail shops in more accessible locations will therefore be able to pay higher rents and the rent-paying ability of retail shops will decline with distance from the point of maximum accessibility. The point of maximum accessibility in a city is taken to be the city centre. The rent-paying ability of retail shops can be expressed as a bid-rent function as shown in ***Figure 6.5***. A bid-rent function is defined as the amount of rent a firm can afford to pay at different locations if profits are to be the same at these different locations.

Commercial firms in offices also require to be close to the city centre, but, since their profitability is not so dependent on accessibility, they have a less steep bid-rent function. And, while households seeking dwellings may desire central city locations in order to reduce the costs of the journey to work and shops, they cannot afford to pay higher rents than retail or commercial firms. The different rent-paying ability of the three land users are depicted in ***Figure 6.5***. Since land is allocated to the highest bidder we find the progression of retail, commercial firms and residential land use outwards from the city centre. Secondary areas of accessibility are formed at the intersection of radial routes and routes from the city centre. This will affect the differing bid-rent

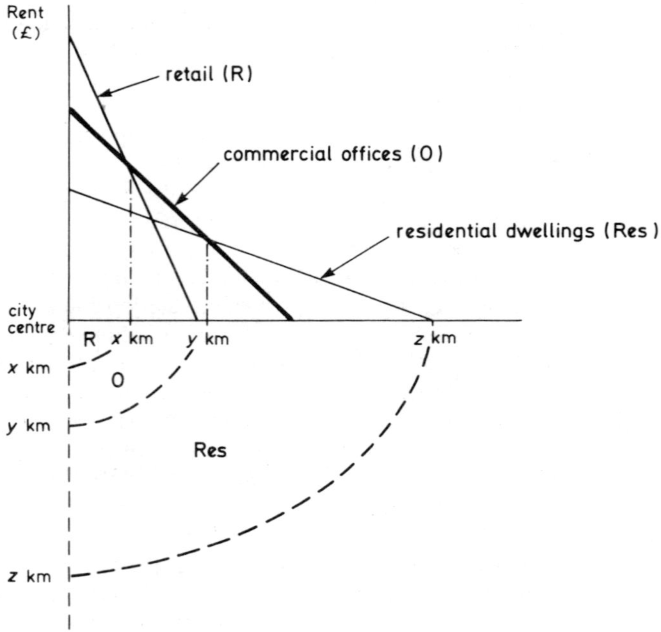

Figure 6.5 Bid-rent functions of three land uses

functions as shown in ***Figure 6.6*** where the point marked A is the position of such an intersection.

The previous discussion suggests that rents for sites decline with distance from the city centre and that retail and commercial firms capture the high rent, accessible locations. These two hypotheses can be verified with reference to the data sources which we have already discussed. Since the gross rateable value can be used as a surrogate measure of land rent, then the rateable value records can be used to identify the nature of the land rent surface in a particular city.

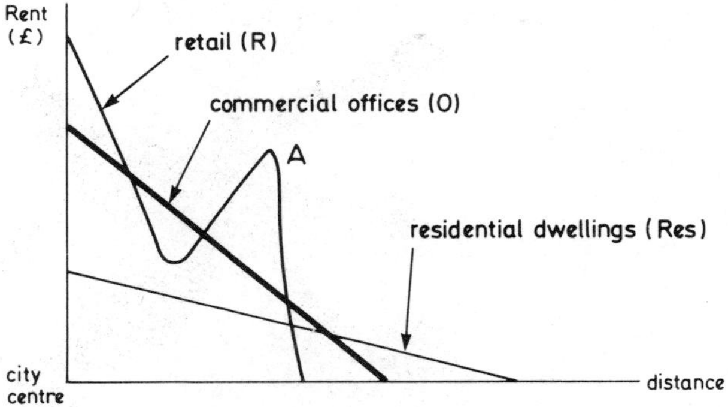

Figure 6.6 Bid-rent functions of three land uses in city with major traffic intersection

To take into account the differences in floor space calculate a ***rate index*** for properties in different locations of the city by dividing the gross rateable value of a property by the area of the ground floor; Herbert (1972) discusses this rate index in greater detail than is possible here. Plot the distribution of the rate index values and compare it with the general distribution shown in ***Figure 6.7.*** In this diagram the highest values are found in the city centre and decrease towards the edge of the city, while secondary peaks are found at major traffic intersections; these secondary peaks also decrease towards the edge of the city.

The general nature of the land use in these high rent areas can be found from an examination of the planning department land use maps and through an analysis of the rateable value records. The rateable value records distinguish between houses, offices and shops. We would expect to find shops and offices in the more accessible areas with high gross rateable values. These records also allow us to determine the number of users in a particular property. Because of the high rents demanded for accessible sites we would expect to find a large number of users in premises located close to the city centre. ***Figure 6.8*** provides a hypothetical example of the differentiation within premises throughout the city which can be easily tested with reference to the rateable value records. In premises close to the city centre shops will tend to be located on the ground floor as a ground

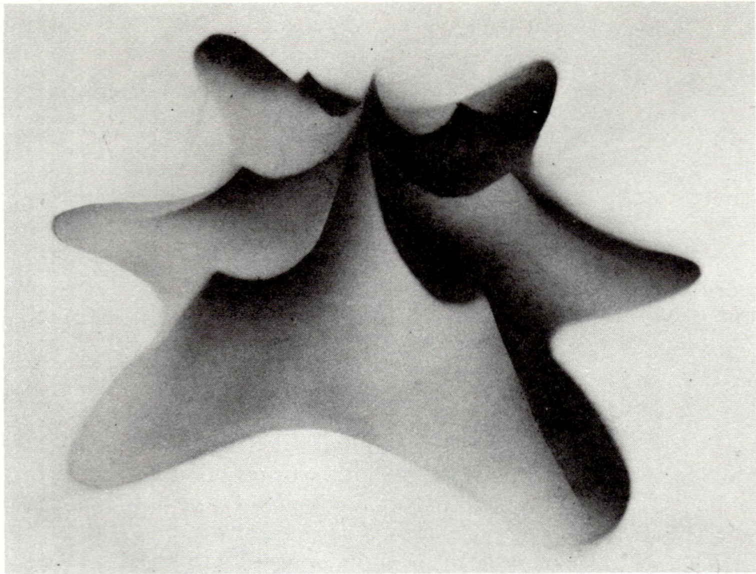

Figure 6.7 Land values in the city

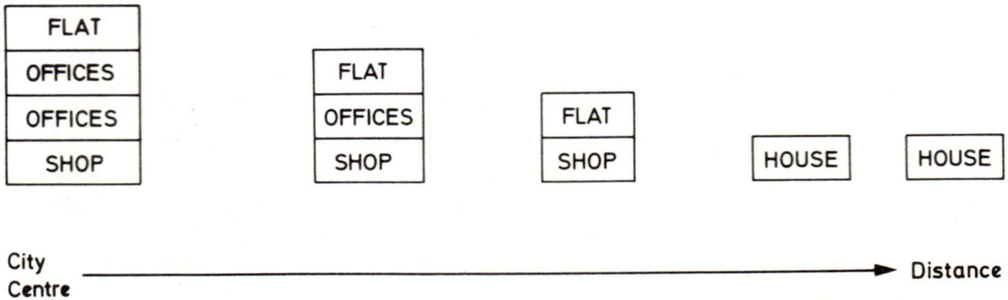

Figure 6.8 Subdivision of premises within the city

floor frontage attracts customers. In the levels above the ground floor we would expect to find offices which require central locations but do not require ground floor premises. Further out from the city centre the number of storeys in premises will tend to decline until the two storey houses in the residential areas begin to dominate the landscape.

6.4.3 The central business district

The central area of cities, commonly referred to as the Central Business District (CBD), is the hub of commercial and retail activity in the city. The CBD can broadly be defined as the area of highest land values in the city. Plot the distribution of the rate index values in your city. By using different cut-off points identify the differing areas subsumed under the title of CBD.

The CBD is not a homogeneous area. Horwood and Boyce (1959) have suggested that the CBD can be considered as consisting of two inter-related parts — the core and the frame. The CBD core is the area of intensive land use with multi-storey dwellings housing offices, retail shops, consumer services, theatres and banks. It has the highest concentration of pedestrian flows and an absence of permanent residential population. The CBD frame is an area of semi-intensive land fringing the CBD core where land use is dominated by warehouses, car sales and services, transport terminals, light manufacturing and some institutional uses.

The CBD and surrounding areas of many large towns and cities in Britain has been the scene of large scale land use changes in the past 10 years. From about 1970 land use developments in these areas was characterised by the construction of large scale office blocks. Ambrose and Colenutt (1975) admirably describe this process in Southwark, London and Brighton and argue that, while such construction profited the property developers, it resulted in a reduction of housing and the enforced break-up of local communities as some residential areas were cleared to make way for office development. If there has been office development in your city read the Ambrose and Colenutt book and use it as a basis on which to study the reasons for, and implications of, recent office development in your city. Ambrose and Colenutt suggest that such studies could be based upon the examination of local authority development plans and contemporary newspaper reports.

6.4.4 The spatial distribution of retail and commercial firms

Inter-urban

The distribution of retail and commercial firms within the urban hierarchy is dependent upon the nature of the goods and services provided by such firms. There are two characteristics of these goods and services which are of particular importance. The ***range*** of a good is the distance over which consumers are prepared to travel in order to purchase the good. The ***threshold*** of a good is the minimum population necessary to support the continued supply of a good. Consider the case of a jeweller's specialising in expensive jewellery and a small grocer's. The jeweller's has a large range — people are prepared to travel far for an expensive bracelet — and a large threshold — it needs to be located close to a large population (***see*** Chapter 2). In contrast, the grocer's has a restricted range — people will not travel far just to buy a tin of beans — and a small threshold. The same characteristics apply to newsagents and cinemas; the newsagent has a limited range and a small threshold while the cinema has an extensive range and a large threshold. The goods and services with extensive ranges and large thresholds are termed ***higher order***, or luxury, goods and services while goods and services with restricted ranges and small thresholds are termed ***lower order***, or convenience, goods and services. The larger the town or city the larger the number of firms and services providing higher goods and services. The hierarchy of towns with respect to population is reflected in the hierarchy of towns with respect to the number of higher order goods and services.

Intra-urban

The relationship between the range and thresholds of the goods and services provided by a firm also determines the location of such firms within the city. Establishments specialising in higher order, luxury, goods will tend to be concentrated in the central area of the city because this is the most accessible point for the majority of the population in the city, while establishments providing lower order, convenience, goods and services will be widely scattered throughout the city.

The predicted spatial patterns of these different types of retail and commercial establishments can be tested by using a number of techniques on the data obtainable from ***Kelly's Directories*** and ***Yellow Pages***. For example assume that you have plotted the distribution of one type of retail establishment as a series of points on a map. If a grid

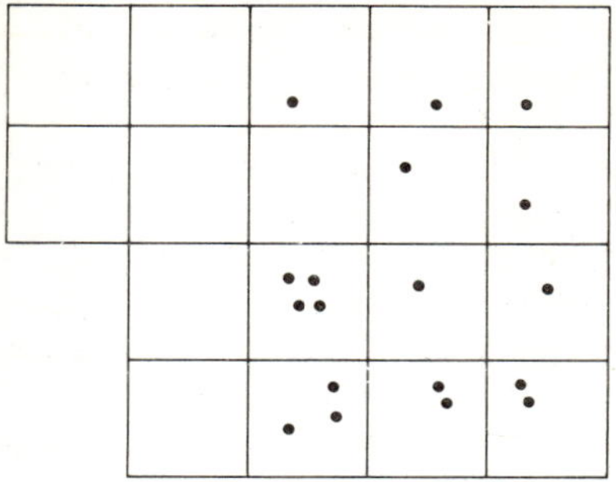

Figure 6.9 Hypothetical distribution of shops

pattern is imposed on the point pattern – achieved by plotting the points with reference to the ordnance survey 1 km squares – then the ***variance–mean ratio*** can be easily calculated. ***Figure 6.9*** provides a hypothetical point pattern of shops and the calculation of the variance–mean ratio for such a point pattern is given in ***Table 6.4***. The higher the value of the variance–mean ratio the more cluttered is the spatial distribution of the points and the closer the value is to 0.0, then the more random is the distribution of points. If we calculated the variance–mean ratio for the distribution of different types of establishments we would expect, from the previous discussion, to find that establishments providing higher order, luxury, goods and services had a higher variance–mean ratio than establishments providing lower order, convenience, goods and services. This hypothesis was confirmed in a study of Reading where Guy (1976) found that the distribution of

TABLE 6.4 CALCULATION OF VARIANCE–MEAN RATIO VALUE (*see Figure 6.9*)

Nos. of shops per cell (X)	Nos. of cells	$(X - \bar{X})$	$(X - \bar{X})^2$	$(X - \bar{X})^2 \times$ nos. of cells
0	7	–1	1	7
1	7	0	0	0
2	2	1	1	2
3	1	2	4	4
4	1	3	9	9
				$\Sigma = 22$

Total nos. of cells = 18
Total nos. of shops = 18

$$\text{Mean of the distribution } (\bar{X}) = \frac{\text{number of shops}}{\text{number of cells}} = \frac{18}{18} = 1.00$$

$$\text{Variance of the distribution} = \frac{\Sigma((X - \bar{X})^2 \times \text{nos. of cells})}{\text{nos. of shops}} = \frac{22}{18} = 1.22$$

$$\text{Variance–mean ratio} = \frac{1.22}{1.00} = 1.22$$

women's clothing shops — a higher order establishment — had a variance–mean ratio of 35.5 while the distribution of sub-post offices — a convenience service — had a variance–mean ratio of 0.59.

The location of certain retail and service establishments will also be influenced by the location of complementary establishments. For example retail outlets profit from the presence of other retail outlets since this concentration and proximity attracts a large number of customers; similarly with some leisure facilities since theatres, cinemas, discos and restaurants profit from the proximity to each other in attracting customers and patrons. In the commercial world the need for face-to-face contact often necessitates the close proximity between certain linked activities and firms such as banks and stockbrokers. The net result is, at least in the larger cities, the emergence of distinct sub-regions within the retail and commercial areas of the city. The 'city' in London is one such sub-region wherein the major banks and other financial institutions are located. The existence of such sub-regions can be verified by an examination of the Goad shopping plans and through an analysis of ***Kelly's Directories*** and ***Yellow Pages***. For example plot the location of insurance offices in your city. Are they clustered together in one distinct area of the city? Does their location correspond to the distribution of bank head offices? When dealing with such spatial distributions it is often useful to discover the centre of gravity of a particular distribution. This can be achieved by calculating the ***mean centre*** of a distribution:

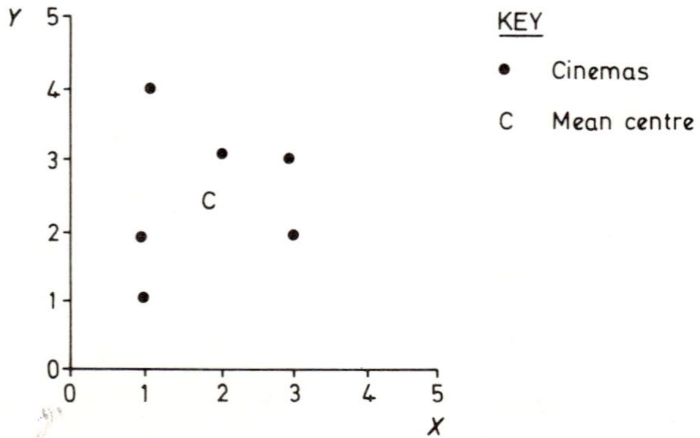

Figure 6.10 Mean centre of distribution of cinemas

assume that the distribution of cinemas in the central city was plotted on a grid square as shown in ***Figure 6.10***. The calculation of the mean centre co-ordinates of this distribution is shown in ***Table 6.5***.

TABLE 6.5 CALCULATION OF MEAN CENTRE (*see Figure 6.10*)

Co-ordinates of mean centre are $\bar{X}$ and $\bar{Y}$

$\bar{X} = \Sigma X_i/N$ where ΣX_i is the sum of all the X co-ordinates of all the points and N is the total number of points

$$= \frac{1 + 1 + 1 + 2 + 3 + 3}{6} = \frac{11}{6} = 1.83$$

$\bar{Y} = \Sigma Y_i/N$ where ΣY_i is the sum of all the Y co-ordinates of all the points and N is the total number of points

$$= \frac{1 + 2 + 2 + 3 + 3 + 4}{6} = \frac{15}{6} = 2.5$$

∴ co-ordinates of mean centre are 1.83, 2.5

The calculation of the mean centre of the spatial distribution of various retail, commercial and service establishments allows us to compare the different distributions and to identify the location of distinct sub-regions within the central city. Since *Kelly's Directories* have been published over a large number of years it would also be possible to use the data obtained in an historical sequence of directories for any one city to plot the changing mean centre of different retail, commercial and service establishments over a long period of time.

6.5 PATTERNS OF OWNERSHIP

6.5.1 Data sources

It is often useful to know the precise patterns of ownership of firms and companies operating within specific towns and cities. Such information is available in a number of

subsidiaries -
Charterhouse SA (France).
Charterhouse Development GmbH (Germany).

† The Charterhouse Group Ltd,

1 Paternoster Row, St. Pauls, London EC4M 7DH (Tel. 01-248 3999)
Telex: 884276
Telgr. Add: CHARVESKO London - EC4
Bank: Lloyds; Midland; National Westminster; Bank of Scotland; National & Grindlays
Office Hours: M-F 9.30-5.30
Loc: Adjacent St. Paul's Underground
Directors: G.N.Mobbs (Ch), G.C.Rowett (MD), P.G.Benham, G.E.Borrett MBE, H.S.Clarke, J.A.Downes MBE, MC, N.G.Foulkes, K.H.Thompson, P.A.H.Viney DFC, M.H.W.Wells
Share Cap: Authorised £35,000,000 Issued £25,782,341
No. Employees: 9,539

Nature of Business

The Charterhouse Group is a public company with financial and industrial interests in the UK and overseas. The Group controls a diversified group of business activities and provides financial and management assistance to encourage the prosperous development of small to medium sized businesses in the UK and abroad.

Principal Operating Subsidiaries:
Development Capital:
Charterhouse Development Limited
London, Paris.
Charterhouse and Associates (Aberdeen) Ltd, Aberdeen.
Charterhouse Canada Limited.
Toronto
Charterhouse Group International Inc., New York, Philadelphia.
Charterhouse Securites Limited, London.
Associate:
Charterhouse Development Capital Ltd
Banking, Pension Fund & Investment:
Charterhouse Japnet Ltd.
London, Aberdeen, Birmingham, Cardiff, Manchester, St. Helier, Geneva, Houston, Nassau. Associates in Frankfurt, New York, Paris, Philadelphia.
Volunteer Holdings Ltd.
Charterhouse Japhet Export Finance.
Charterhouse Japhet Unit Management Ltd
Charterhouse Magna Assurance Co. Ltd
Oil Exploration and Development:
Charterhouse Petroleum Development Ltd.
Insurance Broking:
Glanvill Enthoven & Co. Ltd

Directors: M.H.W.Wells (Ch and Chief Exec), G.C.Rowett (Deputy Ch), P.St.J.Bamford, R.A.Bartholomeusz, G.E.D.Bonham-Carter, N.J.Cosh, N.A.M.Eastwood, B.A.Fireman W.J.Haymes, C.A.Horan, B.C.Johnston, R.S.Napier, P.P.Ralph, D.A.Roberts, C.Rudkin, I.J.Scott, J.H.Sleeman, C.W.Taylor-Young, J.A.Turnbull
Share Cap: Wholly Owned by The Charterhouse Group Ltd
No. Employees: 210
Product Groups: 62

Nature of Business

Provides a full range of merchant banking services, including finance for oil and energy related industries.

Charterhouse Management Courses Ltd,

40 Charterhouse Square, London EC1M 6EA (Tel. 01-606 0121)
Bank: National Westminster, New Oxford Street
Office Hours: M-F 9-5.30
Loc: Adjacent Barbican Underground Station
Directors: J.Grenier, V.Aggett
Product Groups: 83

Nature of Business

Established in 1973 - Now recognised as the leading financial tutors to management in the U.K.
See Display Advertisement near Product table 83.32 in Volume 1.

Chase Products (Engineering) Ltd,

31 Colville Road, Acton, London W3 8BT (Tel. 01-992 1153 and 3570)
Bank: National Westminster
Office Hours: M-F 9-4.30
Loc: 3 minutes South Acton Station off Stanley Road or 6 minutes Acton Town underground down Bollo Lane
Directors: D.Prior (MD), A.E.Prior, R.Cottle
Share Cap: Authorised £5,000 Issued £3,000
No. Employees: 50
Product Groups: 26,35,37,40

Manufacture and Export

Illuminated and non-illuminated signs, hotel entrance canopies, general and precision sheet metal workers.

UK, Australia, Bermuda, New Zealand, Nigeria, Republic of Ireland, South Africa, Spain and Associates throughout U.S.A.

Manufacturing:

Alenco Ltd.
UK, Belgium, France, Italy, Norway, Sweden, The Netherlands, U.S.A.

Charcon Ltd.
UK, The Netherlands

Newage Engineers Ltd.
UK, Australia, Canada, France, West Germany, Italy, Norway, Singapore, Spain, U.S.A.

Kartret Special Panels Ltd.

Kay & Co. (Engineers) Ltd.

Tempest Diesels Ltd

The Harris Plating Co. Ltd.

Distribution and Services:

Edmundson Electrical Ltd
UK

Edmundson Electronic Components Ltd., UK

IDS Distribution Services Ltd.
UK, Saudi Arabia

PTS Edmundson Tools Ltd.
UK

Napcolour Ltd., UK

Spring Grove Services Ltd.
UK Republic of Ireland, The Netherlands

---

† Charterhouse Japhet Ltd,

1 Paternoster Row, St. Pauls, London EC4M 7DH (Tel. 01-248 3999)

Telex: 884276, Foreign Exchange 884306, 884307

Telgr. Add: INTERAXIS - London EC4

Office Hours: M-F 9.15-5.15

Loc: Adjacent St. Pauls Underground

Chatto & Windus Ltd,

40-42 William IV Street, London WC2N 4DF (Tel. 01-836 0127)

Telgr. Add: Bookstore London-WC2N 4DF

Bank: National Westminster, 10 St. Martin's Place, WC2N 4DF

Office Hours: M-F 9-5.30

Loc: Adjacent Strand and Trafalgar Sq. Undgr. Stns

Directors: Mrs N.Smallwood, G.W.Trevelyan, H.Brunner, J.F.Charlton, D.J.Enright, C.Maclehose

Executives: B.A.Strange ACIS (Sec)

Share Cap: Wholly Owned Subsidiary of Chatto, Bodley Head & Jonathan Cape Ltd

No. Employees: 30

Product Groups: 28

Chautard & Co. (Rubber) Ltd,

67 Upper Thames Street, London EC4V 3AH (Tel. 01-248 5711)

Telex: 883771

Telgr. Add: Headmost London-EC4

Code: Bentley Acme 2nd

Bank: National Westminster, Mark Lane, E.C.3

Office Hours: M-F 9.15-5.15

Loc: 2 mins South Mansion House Undgr

Directors: R McC.Cowan, I.A.McInroy, C.B.Roast (Sec)

Share Cap: Authorised £150,000 Issued £150,000

Product Groups: 61

Figure 6.11 Extract from *Kompass* (Volume 2)

A. & H. MELTZER (HOLDINGS) LTD., **TE**
78 Pretoria Road, London N18 1SR
Dorras of London Ltd.
A. & H. Meltzer Ltd.
A. & H. Meltzer (Sales) Ltd.
Meltzer-Prescott Ltd. (*d*)
H. Sims & Co. Ltd.
J. F. Willis (Cinderella Shoes) Ltd.

MELVILLE, DUNDAS & WHITSON LTD., **BU**
21 Blythswood Square, Glasgow G2 4AT
James Y. Keanie Ltd.
M.D.W. Developments Ltd.

MENSWEAR INVESTMENTS LTD., ***Shirley House,*** **TE**
Camden Road, London NW1
Frame Clothing Co. Ltd.
Steel & Glover Ltd.
Steel & Glover (Belfast) Ltd.
Steel & Glover (Taunton) Ltd.

MENTMORE MANUFACTURING CO. LTD., **MS**
Platignum House, Six Hills Ways, Stevenage, Herts. SG1 2AY
D. Leonard & Co. (Highley Pens) Ltd. (A)
Platignum Pen Co. (Australia) Pty. Ltd. — **Australia**
Platignum Pen Co. (New Zealand) Ltd. — **New Zealand**
Colombo Industrial Agencies Ltd. — **Sri Lanka**
Mentmore Manufacturing Co. (S.A.) (Pty.) Ltd. — **South Africa**

JOHN MENZIES (HOLDINGS) LTD., **DI**
Hanover Buildings, Rose Street, Edinburgh EH2 2YQ
Algernon Asprey Ltd.
Jones Yarrell & Co. Ltd.
London Book Match Co. Ltd.
John Menzies (Belfast) Ltd.
John Menzies (London) Ltd.
John Menzies (Transport) Ltd.
Randax-EDP Ltd.
Randax (London) Ltd.
Randax (Scotland) Ltd.
Frank Smythson Ltd. (*d*)
Venture Matches Ltd.

MERCANTILE CREDIT CO. LTD., ***Elizabethan House,*** **FI**
Gt. Queen Street, London WC2B 5DP
The Astley Industrial Trust Ltd.
Astley Acceptances Ltd.
Astley-Gaydon Motor Finance Ltd. (*d*)
R.O.L. Finance Co. Ltd. (*d*)
Astley Leasing Co. Ltd.
Astley Vulcan Finance Ltd. (*d*)
Byrom House Nominees Ltd.
Fred Fishwick (Finance) Ltd. (*d*)
Regent Investment Trust Ltd.
Clearway Electrics Ltd. (*d*)
Dial Contracts Ltd.
Ringwest Motors Ltd.
Perrins Motors Ltd.
Southern Vehicles (Southampton) Ltd.
Standard Railway Wagon Co. Ltd.
Rootes Acceptances Ltd.
Transport Acceptances Ltd. (*d*)
Ulster Acceptances Ltd. *d*)
Northern Ireland Motors Ltd. (*d*)
Totley Investment Co. Ltd. *d*)
Bankers Commercial Security Ltd. (*d*)
Business Credit Ltd. (A)
The Camden Group Ltd.
Camden Motors Ltd.
Auto Buyers (Warren Street) Ltd.
C.G. (Dunstable) Ltd.
Camden Motor Cycles Ltd.
Camden Motor Rentals Ltd.
Camden Motors Fleet Sales Ltd.
Camden Motors (Works) Ltd.
Camford Cars Ltd.
Constellation Publicity Co. Ltd. (*d*)
Continental Car Centre (Mill Hill) Ltd. (*d*)
Finance Acceptances Ltd.
Hillwood Motors Ltd.
E. J. Lee (Leighton Buzzard) Ltd.
U.C.B. (Leighton Buzzard) Ltd.
Chrysler Acceptances Ltd.
Charles Clark & Son Ltd.
T. Cowie Ltd. (A)
Dunelm Trust Ltd. (*d*)

Figure 6.12 Extract from *Who Owns Whom* (Volume 1)

data sources, some of which provide supplementary data on the commercial, industrial and retail structure of urban areas.

Kompass

This is a two-volume publication published each year by Kompass Publishers Ltd., since 1962. There are separate editions for major countries of the world. Volume 1 of the British edition gives the name and address listed in relation to products and services. Volume 2 lists more information on companies listed alphabetically for counties of England, for some individual towns for the whole of Wales and for the whole of Scotland. ***Figure 6.11*** shows an extract from the Greater London section of Volume 2. Note that this source lists the address, nature of business, number of employees and the names of directors. The term 'product groups' refers to a classification used by the publishers, a listing of which is given in Volume 1. Group 28, for example (***see under*** **Chatto and Windus** in ***Figure 6.11***) refers to printing and publishing.

Who Owns Whom

This two-volume directory is published each year by Roskill and Co. (Reports) Ltd. The U.K. edition was first published in 1958. Volume 1 contains a listing by parent company of the names of subsidiary and associate companies. ***Figure 6.12*** is an extract from this volume. The parent company is printed in capital letters while the companies owned by this parent company are listed underneath in alphabetical order. The capital letters to the right of the name of the parent company refer to a classification of activities used by the publishers, for example FJ (***see next to*** **Mercantile Credit Co. Ltd**) indicates banks, financial and discount companies. Volume 2 gives, next to the name of subsidiaries listed in alphabetical order, the name and main activity of the parent company. An extract shown in ***Figure 6.13*** shows that the subsidiaries are listed on the left hand side of the page while the corresponding place on the right hand side gives the name and activity of the parent company.

Company reports

Each registered company in Britain usually publishes an annual report which gives an

Hammond, Hammond & Co. Ltd., *London WC1*	Barrie and Jenkins Ltd.	PA
John Hammond & Co. Ltd., *Newcastle, Staffs.*	J. Compton, Sons & Webb (Holdings) Ltd.	TE
L. Hammond & Co. (Agencies) Ltd., *London EC3*	Hill, Samuel Group Ltd.	FI
L. Hammond & Co. (Australasia) Ltd., *Australia*	Hill, Samuel Group Ltd.	FI
L. Hammond & Co. (Bermuda) Ltd., *Bermuda*	Hill, Samuel Group Ltd.	FI
L. Hammond & Co. (Canada) Ltd., *Canada*	Hill, Samuel Group Ltd.	FI
L. Hammond & Co. (Far East) (1967) Ltd., *Hong Kong*	Hill, Samuel Group Ltd.	FI
L. Hammond & Co. (Holdings) Ltd., *London EC3*	Hill, Samuel Group Ltd.	FI
L. Hammond & Co. (Life and Pensions) Ltd., *London EC3*	Hill, Samuel Group Ltd.	FI
L. Hammond & Co. Ltd.	Hill, Samuel Group Ltd.	FI
L. Hammond & Co. (Reinsurance) Ltd., *London EC3*	Hill, Samuel Group Ltd.	FI
L. Hammond & Co. South Africa (Pty.) Ltd., *South Africa*	Hill, Samuel Group Ltd.	FI
L. Hammond & Co. (Southern) Ltd.	Hill, Samuel Group Ltd.	FI
L. Hammond & Co. (U.K.) Ltd.	Hill, Samuel Group Ltd.	FI
Hammond Lane Industries Ltd., *Eire*	Hammond Holdings Ltd., *Eire*	EN
Hammond Lane (Ironfounders) Ltd.	Hammond Holdings Ltd., *Eire*	EN
Hammond Lane Ironfounders Ltd., *Eire*	Hammond Holdings Ltd., *Eire*	EN
Hammond Lane Metal Co. Ltd., *Eire*	Hammond Holdings Ltd., *Eire*	EN
Hammond Lane Transport Co. Ltd., *Eire*	Hammond Holdings Ltd., *Eire*	EN
Hammond Organ (Rentals) Ltd. (A)	Boosey & Hawkes Ltd.	MS
Hammond Organ S.A. (Pty.) Ltd. (A), *South Africa*	Boosey & Hawkes Ltd.	MS
Hammond Organ (U.K.) Ltd. (A), *Edgware, Middx.*	Boosey & Hawkes Ltd.	MS
Hammond Organ (U.K.) Ltd.	Hammond Corp., *U.S.A.*	MS
R. C. Hammond (Clayton) Ltd., *Crawley, Sussex*	Clayton Engineering Co.	EN
Hammond Refrigeration Ltd., *Eire*	Hammond Holdings Ltd., *Eire*	EN
Trevor Hammond Ltd. (A), *Newport, Mon.*	Joseph Stocks & Sons (Holdings) Ltd.	DI
W. Hammond & Son (Bearded Needles) Ltd., *Loughborough, Leics.*	Sears Holdings Ltd.	IH/EN
Hammond's of Hull and Bridlington	House of Fraser Ltd.	DI
Hammton Investments Pty. Ltd. (A), *Australia*	Berkeley Hambro Property Co. Ltd.	PR
Hammton Properties Pty. Ltd. (A), *Australia*	Berkeley Hambro Property Co. Ltd.	PR
Hamner GmbH., *Germany*	Dunlop Holdings Ltd.	CH
The John Hampden Press Ltd.	Courtaulds Ltd.	TE
Hamplett-Shelton Ltd.	LCP Holdings Ltd.	IH
Hamplett-Shelton (Retail) Ltd.	LCP Holdings Ltd.	IH

Figure 6.13 Extract from *Who Owns Whom* (Volume 2)

Hampo Trading Co. (Pty.) Ltd., *South Africa*	George Weston Holdings Ltd.	FO
E. W. Hampshire (Ireland) Ltd., *Eire*	Reckitt & Colman Ltd.	CH
The Hampshire Furnishing Co. Ltd.	Henderson-Kenton (Holdings) Ltd.	DI
The Hampshire Furnishing Co. (Southern) Ltd.	Henderson-Kenton (Holdings) Ltd.	DI
F. W. Hampshire & Co. Ltd., *Derby*	Reckitt & Colman Ltd.	CH
Hampshire Ignition Ltd.	London Merchant Securities Ltd.	IH
Hampshire Industrial Textiles Ltd.	Youghal Carpets (Holdings) Ltd., *Eire*	TE
Hampshire Mineral Water Co. Ltd.	A. G. Barr & Co. Ltd.	FO
Hampshire Printers Ltd.	Woking Review Ltd.	PA
Hampson Automation Ltd., *West Bromwich, Staffs.*	Hampson Industries Ltd.	EN
H. Hampson (Stretford) Ltd.	Greensquare Properties Ltd.	BU
Hampson Hughes & Co. Ltd., *Manchester*	Haighton & Dewhurst Ltd.	TE
Hampson Jig Tool & Automation Ltd., *Wolverhampton, Staffs.*	Hampson Industries Ltd.	EN
William Hampson (Oldham) Ltd.	Hargreaves, Hamilton & Co. (Engineers) Ltd.	EN
Hampson's (Bolton) Ltd., *Bolton*	Warburtons Ltd.	FO
Hampstead & Highgate Express Printing and Publishing Co. Ltd.	Home Counties Newspapers Ltd.	PA
Hampstead & Surburban Properties Ltd., *London WC2*	Star (Great Britain) Holdings Ltd.	PR
Hampstead Way Investments Ltd., *London WC2*	Daejan Holdings Ltd.	PR
Hampstead Wine Co. Ltd. (A)	Matthews Holdings Ltd.	IH/FO
Hampton & Beebee Ltd., *Wednesbury, Staffs.*	Cooper Industries Ltd.	EN
Hampton & Sons Ltd., *London W8*	John James Group of Cos. Ltd.	FI
Hampton Areas Australia (Pty.) Ltd., *Australia*	Hampton Gold Mining Areas Ltd.	MN
Hampton Bros. (Netherton) Ltd., *Dudley, Worcs.*	The British Bata Shoe Co. Ltd.	TE
C. & J. Hampton Ltd.	Record Ridgeway Ltd.	EN
Hampton Court Finance Ltd., *East Molesey, Surrey*	Park Garage (Molesey) Ltd.	DI
Dick Hampton & Cooper Building Ltd.	Trafalgar House Investments Ltd.	PR
Dick Hampton (Earth Moving) Ltd.	Trafalgar House Investments Ltd.	PR
Dick Hampton (Edinburgh) Ltd.	Trafalgar House Investments Ltd.	PR
Hampton Gold Mining Areas Ltd. (A) (ti), *Australia*	J. H. Vavasseur & Co. Ltd.	FI
Hampton (High Street) Holdings Ltd.	British Industrial Holdings Ltd.	IH/PR

account of the year's trading figures. Other information included in the report varies from company to company but most of them give precise details of shareholdings, the names of directors and major shareholders and a general summary of the company's subsidiaries and associate companies. These reports are easily obtained by writing to the companies most of which will send you a free copy if you include a stamped addressed envelope. As a last resort the company reports should be available in Companies House in either London or Cardiff.

Directory of Directors
This annual directory was first published in 1879. The 1978 edition lists the names and addresses of over 44 000 men and women who hold directorships in one or more of the principal public or private companies in the U.K. Under the name of a director — listed in alphabetical order — the directory lists the name of all companies in which the person holds a directorship.

Who's Who
This directory, first published in 1848, is now a major British institution, an entry in which indicates having arrived or been born into the British establishment. The directory lists individuals by alphabetical order of surname and gives a basic summary of their place of birth, family background, education, interests and hobbies.

6.5.2 Outline of a project

The data sources which we have just considered, all of which should be in your main public library, provide a mine of information. In this section we will outline only one of the many possible shafts which can be sunk into this rich vein of data.

Assume that you are interested in identifying the patterns of ownership for the firms and companies in your town and city. This is a major undertaking and is best tackled as a group project. The basic steps to be undertaken are shown in ***Table 6.6***. Begin by making a list of the most important firms and companies in your town from the basis of your own knowledge and the information provided by ***Yellow Pages***, ***Kelly's Directory*** and ***Kompass***. Check if these companies are subsidiaries of parent companies by referring to Volume 2

TABLE 6.6 DATA SOURCES FOR A RESEARCH PROJECT

Step	Relevant data sources
Identification of companies	*Kompass, Kelly's Directory, Yellow Pages*
Tracing parent company	*Who Owns Whom* (Volume 2)
Character of parent company	*Who Owns Whom* (Volume 1)
Name of directors and major shareholders	*Company Reports, Kompass*
Other interests of directors and major shareholders	*Directory of Directors, Who's Who*

of ***Who Owns Whom***. The character of these parent companies can in turn be discovered with reference to the relevant entry in Volume 1 of ***Who Owns Whom*** and ***Kompass***. More detailed information may be available in the company reports which, if they are not available in the public library, may be obtained by writing to the company headquarters. These reports as well as ***Kompass*** provide the names of directors and major shareholders. By using the ***Directory of Directors*** and ***Who's Who*** the other financial interests and background information of these people can be ascertained.

CHAPTER 7 DATA SOURCES: A POSTSCRIPT

7.1 INTRODUCTION

In the previous chapters data sources and statistical techniques have been introduced around a number of themes. However, the previous discussions do not exhaust the range of data sources available for the analysis of towns and cities. In this chapter a number of other relevant data sources will be discussed; these sources are not presented in any thematic form.

7.2 ANNUAL ABSTRACT OF GREATER LONDON STATISTICS

Since 1966 the Greater London Council (GLC) have published an annual abstract of statistics relating to the boroughs which make up the GLC. This provides a comprehensive source of information covering a variety of topics. The information is provided in a series of tables covering 12 sections (*Table 7.1*). In general, the respective figure for different variables are given for each borough with separate totals for inner, outer and all boroughs. Where appropriate, the respective figures for England and Wales are also

TABLE 7.1 TOPICS COVERED BY GLC ANNUAL ABSTRACTS

Section	Topic(s)
1	General information, elections and weather
2	Population and vital statistics
3	Economic activities
4	Land use
5	Transportation and communications
6	Health and social services
7	Education
8	Housing
9	Environmental services
10	Protective services: ambulance, fire, police, licensing and regulative functions
11	Amenity and cultural services
12	Local government finance

TABLE 7.2 OPEN SPACE AND PUBLIC LIBRARIES IN LONDON BOROUGHS, 1972 (From *GLC Abstracts*, 1972, Volume 7)

Boroughs*	Estimated total population (1972)	Hectares of open space	Number of public libraries
City of London (C)	4320	–	7
Barking (16)	158 660	296	12
Barnet (30)	305 580	806	16
Bexley (18)	217 860	641	14
Brent (28)	274 760	305	12
Bromley (19)	306 290	946	14
Camden (2)	196 650	347	14
Croydon (20)	334 000	1180	15
Ealing (27)	299 270	499	13
Enfield (32)	266 020	784	15
Greenwich (6)	216 270	550	14
Hackney (4)	215 040	286	19
Hammersmith (11)	181 860	153	8
Haringey (31)	235 570	362	11
Harrow (29)	203 390	385	10
Havering (15)	246 940	615	10
Hillingdon (26)	236 420	968	15
Hounslow (25)	206 720	374	10
Islington (3)	194 200	35	11
Kensington & Chelsea (12)	182 190	42	6
Kingston upon Thames (23)	138 930	229	8
Lambeth (9)	300 250	213	13
Lewisham (7)	269 900	314	18
Merton (22)	176 880	714	9
Newham (17)	231 580	157	11
Redbridge (14)	237 630	720	12
Richmond upon Thames (24)	172 730	1989	12
Southwark (8)	252 960	171	18
Sutton (21)	169 200	286	9
Tower Hamlets (5)	158 780	126	15
Waltham Forest (13)	233 200	588	13
Wandsworth (10)	296 390	803	13
City of Westminster (1)	231 380	460	16
Greater London	7 344 820	16 343	413

*Figures in brackets relate to ***Figure 7.1***

Figure 7.1 Boroughs in the Greater London Council

given for comparative purposes. It is no exaggeration to say that this provides the most comprehensive source of information for any one particular city. The range of information is enormous covering, for example, the number of legitimate live births in each borough, the number of public houses in each borough and the number of pupils in each borough who have received a full school medical examination. The abstracts also include a map of the borough boundaries and a series of maps which indicate the boundaries of gas regions, postal districts, health service regions and even the areas covered by the different sewage and sludge disposal works.

This enormous range of data can be used for a multitude of purposes. As with the small area statistics of the Census the data can be used to test various models of urban structure and to identify the nature of the residential mosaic (***see*** Chapter 3). The abstracts can also be used to note the spatial variation of a variety of phenomena. For example we may wish to consider aspects of the ***quality of life*** in the different GLC boroughs. The quality of life is a difficult concept to measure; however, we can suggest that an important part of the quality of life is access to open space and public libraries. These two characteristics can be indirectly measured by recourse to the annual abstracts which give the hectares of open space and the number of municipal libraries in each borough. The data for 1972 are presented in ***Table 7.2*** while the borough boundaries are shown in ***Figure 7.1*** For each borough, calculate the hectares of open space and the number of public libraries per 1000 of the population. Rank the results and comment on the patterns you discover. Is there any difference between the inner and outer boroughs? What other indices would you use to measure the quality of life?

7.3 THE LOCAL AUTHORITY AS DATA BANK

Throughout this book reference has been made to the types of information available from the local authority. For the study of any particular city the local authority can be considered as the most important data bank. Firstly, we can draw upon the information provided by the various development and structure plans (***see*** Chapter 6). Secondly, there are the data provided by the various land use transportation studies undertaken by many major local authorities since the mid-1960s. The reports of these studies provide a welter of empirical material concerning car ownership, land use arrangements and traffic flows.

Thirdly, there are the various types of information held by the different departments of the local authority. The local authority housing department, for example, may be able to provide information on local authority mortgages, house improvement grants and council house allocations and transfers. In many local authorities the data collected and maintained by the different departments may be held by the planning research department or its local equivalent. Such departments often publish the results of an analysis of such data in a series of research reports. Before starting a study of one particular facet of your local town or city check first with your local planning department to see if the appropriate data have not already been collected and analysed.

7.4 THE RENT REGISTER

Since 1966 registered rents in the private rented sector of the housing market have been fixed by a rent officer who is independent of either central or local government. Landlords and/or tenants may apply to the rent officer for the rent of a property to be fixed at a fair rent level. A 'fair' rent is assessed on the assumption that supply and demand for rented acommodation are equal; a fair rent is the likely market rent assuming that there is no scarcity of rented accommodation. The applications for fair rent registration are filed at the Rent Officer's Office in the rent register which is freely available for public inspection. Each application form gives the name and address of the landlord and their agent; the name of the tenant(s); the address, description and rateable value of the property and the level of rent which has been fixed by the rent officer. Although this source does not cover all properties in the private rented sector, it does provide a useful source of information on landlords and levels of fair rent. The rent registers can, therefore, be used to identify a sample of landlords and to examine the spatial distribution of fair rent levels throughout a city.

7.5 NEWSPAPERS

The local newspaper provides a cheap and handy source of information about various aspects of particular cities. We have already noted in Chapter 2 that the local newspaper can be used to demarcate the spheres of influence of particular towns. There are several other ways in which a local newspaper can be used. First, the local newspapers invariably have a large advertising section devoted to housing for sale. A cursory glance at such a section will reveal that each advertisement is preceded by an area name identifier. In many cases these names will cover a very wide area of the city, but they can still be used to extend our knowledge of the broad structure of the private housing market. By noting the average selling price and the type of properties advertised for different areas of the city we can build an up-to-date picture of the housing element of the residential mosaic. This form of analysis provides a useful, if somewhat crude, extension to our knowledge of a particular city.

Secondly, most, if not all, local newspapers provide a coverage of all important planning decisions reached by the local authority and sometimes an analysis of the kinds of conflicts involved in such decisions. Future road building, housing, welfare and recreation schemes are regularly covered by local newspaper reports, often with a survey of the opinions of those people most affected by such schemes. These newspaper reports include a wealth of background information. You may find it useful, in this respect, to start a file, based on relevant newspaper clippings, on topics which most interest you. ***Figure 7.2*** provides an example, in ***collage*** form, of newspaper clippings taken from the ***Bristol Evening Post.***

There are some national newspapers which provide an important source of general information on urban matters. ***The Times***, ***Financial Times***, ***Guardian*** and ***Daily Telegraph*** all contain occasional, and often regular, reports about major planning proposals, housing policies and other issues relevant to an understanding of modern urban life in this country. The ***collage*** in ***Figure 7.3*** contains a number of clippings taken from national newspapers concerned with various aspects of housing and housing policy. You will find it very useful as a source of reference to start a series of files composed of national newspaper clippings suitably arranged around a number of themes.

7.6 BACKGROUND STATISTICS

The sole use of aggregate spatial statistics, as in the ward census data, makes it difficult to comment on the relationship between a series of attributes at the level of individual households. While the ward data allow us to comment on the relationship between tenure and socio-economic status for large population groups in specific areas, they do not allow

The worm turns. A city's downtown residents have revolted against decisions to move them out and redevelop. Robert Waterhouse reports

ST PAUL'S IS a flyblown inner city district of Bristol where a large West Indian community flourishes beside immigrants from most other parts of the world and refugees from the rest of Britain. Crowded into dusty two and three storey terraces, overshadowed by Bristol's soaring office towers and harrassed by through traffic, St Paul's residents suffer authentic urban stress. But the latest threat — demolition and redevelopment — has caused more bewilderment and anger than relief. The last thing many residents want is to be moved from their no-questions-asked bedsits.

At a public meeting last week some four hundred tenants and interested parties turned out to face the council's top brass. Undeterred by the presence on the platform of the housing committee chairman, Councillor Graham Robertson, the planning committee chairman, Councillor Brian Richards, and Cyril Fortune, director of housing, residents complained that they wanted improvement not demolition (of which there is plenty in the area) and demanded proper consultation before decisions were taken. Owner-occupiers doubted whether they would receive enough compensation to allow them to buy houses elsewhere and, said they had already been approached by agents representing the council. The upshot was that Councillor Robertson promised to review redevelopment plans and to hold further meetings. Bristol's first major attempt at public participation in the rehousing process ended in stalemate and frustration all round.

Bristol council officials, carrying on routine slum clearance techniques, have met resistance where they seemed least likely to — from a drifting population living uneasily with the law. Other more established working class communities like Totterdown and Easton were erased with hardly a whisper. The reason, of course, for this strange turn of events is a new awareness of people power backed by the organisation of a pressure group who talk the planners' language but wholly mistrust their methods. SPAM, or the St Paul's and Montpelier Action Group was basically formed to fight the many road proposals which haunt inner Bristol but it has found, in St Paul's redevelopment, an even hotter issue.

SPAM's problem is that it cannot claim to represent residents of the five or six streets between City Road and Grosvenor Road threatened with demolition. Most of its leading members live in Montpelier, the pleasantly mixed district north of St Paul's which has the advantage of a steep hillside to provide hidden gardens and splendid views. Not even Roy De Freitas, the West Indian chairman of the St Paul's Community Council, lives in the threatened survey area. So the first priority of a successful campaign has to be a residents' committee which will take its own initiative.

Bill Reynolds, a schoolteacher who speaks for SPAM and lives safely out of the way on a Montpelier ridge, makes no suggestion that his organisation knows best. On the other hand he is unquotably critical of local authority methods of reshaping inner city life — not only its housing but its roads and office development policies. Bristol district council (Labour-controlled) is already at odds with its Tory-ruled big brother, Avon County Council, over the provision of office space in the city. Present estimates are that there will be over 10 million square feet of office space in the city by 1978, of which a sizeable proportion could be empty. Following a squatting exercise which attracted maximum publicity, Bristol council has proposed a two-year moratorium on office building.

Long-term road plans are increasingly in doubt, and await the publication of the land use and transportation study at the end of the year. But while Montpelier stays theoretically threatened by the Outer Circuit Road, Saint Paul's faces the immediate and certain danger of carrying increased traffic when the second stage of the M32 Parkway opens. This will terminate at a roundabout just east of the district, from where there are £500,000 proposals to "improve" City Road and Ashley Road. It's difficult, on such tight-packed streets, to see how improvement could be carried out without demolition — so the roads and housing issues are inextricably linked.

Moreover, housing along Newfoundland Road has already been demolished in preparation for the third stage of the Parkway.

It is becoming increasingly obvious that Bristol's unchallenged position as capital of the South-west (significantly it is nearer to London than to Truro) is creating as many problems as opportunities. There is an awful lot of hard cash wanting to invest in the city, and that is inevitably driving people out. Conditions in inner residential areas quite as bad as anything in rundown industrial cities persist in the face of — and in contrast to — such wealth. If it is natural to wish to clean up St Paul's, the net result could be to drive some residents outside society. Bad for the image, but even worse for the frail harmony that maintains today.

Families plan brings storm of protest

By NIGEL HEATH

A plan to house problem families close to luxury homes at Redcliffe Bay, near Portishead, raised a storm of protest this afternoon.

The site at Haven's Field is one of five in the Weston-super-Mare, Nailsea and Portishead area being considered b yWoodspring council.

They want a suitable place to park mobile homes for council tenants whose disruptive and anti-social behaviour make them a nuisance to their neighbours.

But Redcliffe Bay residents, whose £30,000-plus homes are near the steep sloping wooded site, are hotly opposed to the use of the neighbourhood.

"This type of dtvelopment is not suitable for a quiet residential area," said parish councillor Alan Pickering.

He said he was horrified at the proposal.

Use of the site next to Redcliffe Bay's only open space is also worrying nearby council tenants.

The Weston sites being considered are at a former caravan park off Locking Road, land next to the police transport department at Bournville and a site near the railway line behind Williton Crescent. The land being considered at Nailsea is the farmyard of Elm Farm.

Woodspring housing director Mr Don Hardy said today he had also received protests concerning these sites. All five would be considered by the council s housing committee tomorrow night.

Public Notices 5

City of Bristol COMPULSORY PURCHASE OF LAND in St. Paul's

Notice is hereby given that the City Council of Bristol have made the City of Bristol (No. 5 Gwyn Street, St Paul's) Compulsory Purchase Order 1978 under the Housing Act 1974. They are about to submit this Order to the Secretary of State for the Environment for confirmation and if confirmed the Order will authorise them to purchase compulsorily the land described below for the purpose of securing or assisting in securing all or any of the objectives specified in paragraphs (a) to (c) of sub-section (2) of Section 36 of the Housing Act 1974.

A copy of the Order and of the accompanying map may be seen at all reasonable hours at the offices of the City Clerk, Room 360, Third Floor, The Council House, College Green, Bristol, BS1 5TR.

Any objection to the Order must be made in writing to the Secretary of State, Department of the Environment, Froomsgate House, Rupert Street, Bristol BS1 2QN before the 21st day of July 1978, and should state the title of the order and the grounds of objection.

DESCRIPTION OF LAND

130 sq metres or thereabouts, void dwelling house together with rear garden, thereto 5, Gwyn Street, St Paul's, Bristol.

Dated this 28th day of June, 1978

JOHN A. BROWN
City Clerk

The Council House,
College Green,
Bristol BS1 5TR.

It's a carve-up, complain city council Libs

By JIM WEBBER

Liberals on Bristol City Council today protested about a "Con-Lab carve up" of committee seats.

Cllr Gillian Beedell, one of the three Liberals representing Cabot Ward, told the Evening Post that they were considering action over allocation of seats at today's annual meeting.

"Liberals are angry that they have again been allocated only three places on the same committees and without consultation," she said.

"The system is unfair and undemocratic."

Cllr Beedell said Labour took two-thirds of the 165 committee seats. Conservatives chose what they wanted from the rest and Liberals were left to fill the gaps.

Anger at site grows

By NIGEL HEATH

Sub-contractors outwitted angry residents in a big development battle at Pill, near Bristol, today.

Housewife Mrs Margaret Norman parked her car to make it difficult for lorries to get into the Redland Housing Association site at Crockerne Drive. But two workmen picked it up by the back wheels and moved it out of the way.

Residents are protesting over the dangerous access to the site and say the three-storey block of 15 flats is too big for the area.

But sub contractor Mr Barry Rubery said residents should have got their problems sorted out before he moved on to the site.

"I moved the car because it was creating a danger," he said.

Mr Rubery said he made an alternative access to nearby farmland so that his earth moving lorry would not have to use the narrow road.

Local police kept an eye on the angry villagers, who had draped a 40ft long banner across fencing on the site.

Father of two Mr Carl Harrell said they had told Woodspring two years ago that the small plot near their homes was not suitable for the flats.

Groups get £140,000 windfall

By DENNIS LEHANE

BRISTOL is to get £140,000 aid from the Government's urban aid programme, it was announced at Westminster yesterday. It is part of a £10 million scheme to assist towns and cities with special social needs.

Bristol received aid for six projects.

Windmill Hill city farm: almost £16,000; the Housing Advisory Centre at Montpelier: almost £38,000; Bristol Play Bus Project: £8,300.

The Barton Group for deprived and handicapped children: £20,000; the sponsored childminding scheme: more than £9,000; and a new community social worker will cost more than £12,000.

Full details of the grants' use are not yet known, said Bristol City Council spokeswoman Mary Wright.

Young face a bleak future

Nailsea lacks land for industrial purposes and teenagers have a bleak future in looking for local employment a councillor has claimed.

Cllr Roger Harris, chairman of Woodspring's economic development sub-committee, told councillors at Weston-super-Mare yesterday: "We cannot see where the school leavers of Nailsea are going to be employed in five years time. Within the town there is a record of unemployment."

It was agreed to defer consideration of use of land at Nailsea on the old sewage works site on which it had been suggested £1,000 should be spent in providing a skateboard area.

MP acts over bus threat

A THREAT of another fares rise on Bristol city bus services today brought backing from Kingswood Labour Mp Mr Terry Walker for the idea of Government financial aid.

Mr Walker said the special position of Bristol's services, and the further burden on the community, called for a joint approach by the district councils and Avon.

Figure 7.2 Local newspapers as a data base

HOUSING POLICY

Closing the great housing gap

How can we solve the housing dilemma by 1994? Andrew Gilg, in the first of two articles, argues the case for making the house building programme the key economic regulator in the Budget

IN TWO articles last September I examined the factors that lay behind the explosion of house prices during the years 1971 to 1973, and on this basis I went on to examine the numbers of houses that we would need to construct for the next 20 years. I chose 20 years because it is the time span during which the children living today will have to be housed, and also because it is the most optimistic time needed to solve the problem of today's three million slum households.

The important factors in 1971 to 1973 were found to be mortgage availability (a twofold increase in mortgage loans), population structure (the entry into the housing market of the post-war population bulge), and the rate of housebuilding (a steadily decreasing rate of housebuilding particularly in the public sector). Other factors such as land prices and planning controls were found to be symptoms rather than causes.

Since last September events in the housing market have confirmed these conclusions. Prices have drifted downwards in face of a much reduced flow of mortgage finance, and consequently private house building has declined even further to a predicted rate unequalled since the slump of 1929.

The one factor that remains constant and enduring is the growing number of people need-

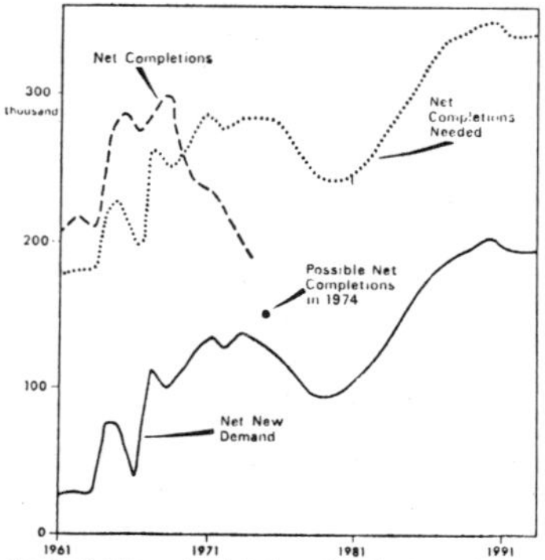

Graph of net new housing demand, and net completions needed for England and Wales up to 1991

that an assured total sum of cash should be allocated in the Budget each year in light of the projected demands for housing in the near future. According to the financial position of the country at the time other projects would be allocated the sum remaining, so that the cycle of lean and fat years in housing would be suppressed and the cyclical effects passed on to less important sectors of the economy.

This radical reappraisal of the role of housing in the national economy would remove the first drawback to the forecasting method shown in the diagram.

The second drawback concerns the localised nature of the housebuilding industry. Housing problems are not general over the whole country, and many towns do not have a housing problem. The problem areas are the inner-city areas, growth points such as oil in Aberdeen and the rural fringes of the major cities. The problem is then one of allocating the national targets to specific geographical locations.

The theoretical solution to this problem is already known and used by local planners in

court proposal

Tenants suffer in unrepaired homes

By Gareth Parry

Striking evidence of widespread hardship caused by fail-

to the landlord has been out of order for over a year. There is no light in the toilet, bathroom.

that many tenants have great fear of eviction and will suffer dreadful conditions of repair.

The CAB survey shows there is a wealth

carry out repairs i
paper published t
National Associatic
Advice Bureaux.

The paper, whi
sent to the Rent
Committee at the
of Environment, is
a survey over the
period of April-Ju
CAB throughout
were asked to mo
this kind — a to
inquiries (17.5 pe
CAB inquiries)
ended March 31. i
cerned with hous
and land.

Replies from
give example
cases of sever
by the landlo
activity to ca
repairs despit
obligation to
clearly illust
conditions in
ately-renting

"Tenant ha
bathroom as
little, or it was
buyers market,
building strike,
cost yardstick v
too low.

It is this com
the housing ind
building trades
gage companies
in nineteenth
and practices, t
with the mixed
present that n
thing has brou
present near d
tion. If the cur
pletions is ever
curve of ne
needed, and if t
are to be buil
places, dramatic

Squeeze likely to hit council housing hard

By JUDY HILLMAN, Planning Correspondent

Investment in public housing next year may be drastically reduced as part of the Chancellor's package of expenditure cuts to satisfy the International Monetary Fund, it was learned yesterday.

tion, apart from the need to offset the effect of rising interest rates on existing holdings. Municipalisation provides

mised necessary legislation for the change to housing investment programmes. This would mean that a council could switch up to 25 per cent of the money allocated for new building, for example, to improve-

The Financial Times Tuesday April 23 1974

HOME NEW

Crosland tells councils to buy unsold private houses

BY JOE RENNISON

A SERIES of policy directives to help promote an immediate increase in local authority housing programmes was announced yesterday by Mr. Anthony Crosland, Secretary for the Environment.

The proposals, contained in a circular to local authorities in England and Wales, complement the policies laid out in the

should be that recommended by the district valuer. Normally local authorities should stick to moderately priced houses for buying, but this obviously would vary from region to region.

Local authorities will also be enabled to attract the services of private builders who, at present, have land for development for the private sector but

see the bulk of that money used by the local authorities to increase substantially the number of homes available to rent.

Local authorities should not sell their houses indiscriminately, since it is generally wrong where there is a strong need for rented homes. There may, however, be areas where the sale of council houses is

provide
e.
ould in
uild for
they can
eeting a
sale and
act from
providing
The ban
labour
ling for

Flats go out as Act comes in with a pessimistic note

The first day of operation of the new Rent Act, which gives security of tenure to most furnished tenants, and for the first time entitles them to apply to a rent officer for the fixing of a fair rent, brought a spate of complaints from estate agents that landlords were witholding such accommodation from the market.

The process has been going on since March, many agents reported, when the Government announced its intention of bringing this legislation on to the Statute Book. It has consisted, not so much of evictions of existing tenants, as of holding empty flats and houses off the market once tenants have left.

more helpful two or three years ago, during the boom days of property speculation when furnished tenants

One section of the market which has apparently been little affected so far by the new law is the high-price luxury

Minister moves to head off housing cash protest

By JUDY HILLMAN, Planning Correspondent

Mr Reg Freeson, Housing and construction Minister, has written personal letters to many London borough council leaders to try to head off angry recriminations over the shortfall in cash allocations for housing programmes

Although London is getting about £155 millions of the £432

Government lopped £50 millions off municipalisation, the programme under which the Labour Party intended to take-over the private rented sector. London councils have been particularly active in this field. Their recent bids for next year came out higher than the £64

at current levels of activity, which are very low.

The new cash limits mark an end to local authorities, freedom to be as active as they wish with their programmes, provided they stick to White-

virement. But there is still no scope for switching money from new building into improvement, whether public or private, for which many councils have campaigned. This procedure is promised in the housing

the present 12 months. Comparable rehabilitation figures suggest 12,780 approvals over 15 months, compared with 10,925.

Wales will be much the same as this year with 500 improvements and 1,000 homes from new building, while Scotland is

Shelter said that there should be some form of housing allowrance for everyone, whether the rent or buy.

Although it is known that Mr Anthony Crosland's unpublished housing finance review carried out a detailed inquiry

Figure 7.3 National newspapers as a data base

us to comment on the relationship between these two attributes at the level of individual households. To avoid the dangers of the ecological fallacy it is necessary to either conduct questionnaire studies or to obtain more suitable data sources. Fortunately, there are three relevant data sources which are available.

7.6.1 The General Household Survey

The General Household Survey, which is conducted by the Office of Population Censuses and Surveys, is based on a sample of approximately 15 000 households in Great Britain. The reports of this survey have been published on an annual basis since 1971. The reports cover six main topics – housing, internal migration, employment, education, health and health services, and fertility. For example, *Table 7.3* lists the distribution of socio-economic

TABLE 7.3 DISTRIBUTION OF SOCIO-ECONOMIC GROUPS BY TENURE CATEGORIES (From General Housing Survey, 1976)

Socio-economic groups*	Tenure (%)					
	Owner-occupiers (owned outright)	Owner-occupiers (with mortgage)	Renting from local authority	Renting unfurnished	Renting furnished	Total
3,4	25	66	2	3	3	99
1,2	34	48	11	5	2	100
5	28	44	15	8	5	100
6	30	26	29	11	4	100
8,9	19	29	42	8	1	99
7,10	18	15	53	12	2	100
11	16	5	65	12	2	100

**See Table 3.4* Non-manual groups are 1,2,3,4,5 and 6.

groups by tenure categories as shown by this survey. Note that the distribution of non-manual workers is biased towards owner-occupation whereas the distribution of manual workers by tenure categories is biased towards renting from the local authority.

7.6.2 Family Expenditure Survey

This survey was introduced in 1957. It is based on an annual sample of approximately 11 000 households in Great Britain. The reports of this survey, which are published

TABLE 7.4 DISTRIBUTION OF INCOME CATEGORIES BY TENURE TYPE (From Family Expenditure Survey, 1976)

Tenure categories	Average weekly income						Total
	Under £25	£25–£50	£50–£100	£100–£150	£150–£200	+£200	
Renting from local authority	17.8	20.6	39.7	15.9	4.1	1.9	100
Renting-unfurnished	20.9	29.0	32.7	13.4	2.7	1.3	100
Renting-furnished	15.6	24.5	42.0	13.6	2.8	1.5	100
Owner-occupation	5.4	15.8	38.1	26.8	8.8	5.1	100

annually, list the pattern of expenditure by various income groups. A variety of types of expenditure are covered, including payments for housing, electricity and various types of consumer durables. The reports also list the income distribution of the different tenure categories. ***Table 7.4***, which is based on data taken from the 1976 Family Expenditure Survey, shows how those in owner-occupation tend to have higher weekly incomes than those either renting from the local authority or renting from a private landlord.

7.6.3 The New Earnings Survey

This annual survey, which is conducted by the Department of Employment, consists of a sample of approximately 170 000 employees in Great Britain for one pay period in April of each year. The reports of this survey include information on the average weekly earnings of workers stratified by age, occupation and industrial classification. Other topics covered by these reports include labour turnover and annual entitlement to holiday with pay. The information is also available on a regional basis. Weekly earnings are also given for the broad categories of non-manual and manual workers. ***Table 7.5***, for example, lists the average gross weekly earnings of manual and non-manual workers by different age categories. These data are represented in ***Figure 7.4***. Note how young manual workers earn higher incomes than young non-manual workers, but as one progresses through the age categories the income of non-manual workers tends to increase more than the income of manual workers. While non-manual workers have an income peak between 40 and 49,

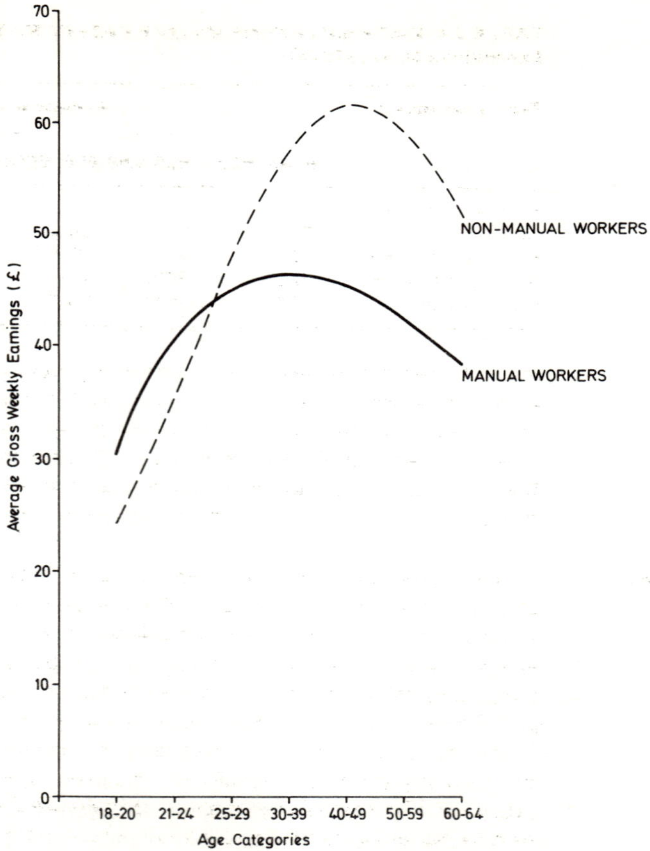

Figure 7.4 Earnings by age of occupational categories

TABLE 7.5 WEEKLY EARNINGS BY AGE AND OCCUPATIONAL CATEGORY (From New Earnings Survey, 1972)

	Gross weekly earnings (£)	
Age category	Manual workers	Non-manual workers
18–20	30.7	24.4
21–24	40.7	35.6
25–29	44.7	47.6
30–39	46.4	57.7
40–49	45.2	61.6
50–59	42.4	58.9
60–64	38.5	51.9

manual workers' peak is at 30–39. These differences in income throughout the life cycle are important to our understanding of the tenure split between the two occupational categories. Since non-manual workers tend to have steadily rising incomes throughout the life cycle up to 49 years of age they have a better chance of obtaining mortgages, which invariably have a 20–25-year repayment period, and thus their chances of entering the owner-occupier sector are better than manual workers whose incomes tend to decline much earlier in life.

These three data sources, which are of considerable value in themselves, provide a wealth of information which can be used to complement aggregate spatial statistics. In this particular case figures have been taken from the three sources which shed further light on the finding that, at the level of ward data, there is a relationship between tenure category and occupational status.

7.7 SOURCES ON SOURCES

There are a variety of books and monographs which provide information on where certain forms of data may be obtained. Perhaps the most comprehensive coverage is available in the ***Guide to Official Statistics*** which was published by the Government Statistical Service in 1976. This Guide covers all official, and the more important non-official, sources of statistics for the United Kingdom. It also includes a listing of all relevant publications which contain a significant amount of statistical information, e.g. Command Papers, Select Committee reports and the publications of various institutes and professional bodies. For the official sources the Guide gives the scale, time series and the nature of the statistics.

Sources of Economic and Business Statistics (Edwards, 1972) and ***Sources of Social Statistics*** (Edwards, 1974) provide an effective and critical guide to the range of official statistical sources. A more comprehensive coverage is given in the four-volume ***Reviews of United Kingdom Statistical Sources*** published jointly by the Royal Statistical Society and the Social Science Research Council. These volumes provide references to the main sources of official and non-official data sources. The availability and limitations of the respective sources are also discussed. Each volume of this series covers two topics: Volume 1 covers personal social services and voluntary organisations in the personal social services field;

Volume 2 is concerned with central government routine health services and social security statistics; Volume 3 is devoted to housing in Great Britain and Northern Ireland, while Volume 4 covers leisure and tourism.

7.8 GENERATING YOUR OWN DATA

If you have consulted all the relevant data sources and find that the data you require is not available then you may feel that you will have to generate your own data. This can be a costly and time-consuming business. Before embarking on any fieldwork or questionnaire exercises, make absolutely sure that the required data are not available. If you have checked, check again; make sure that all possible, and not so possible, data sources have been consulted. If you still find that the data are not available then you may have to embark on some form of fieldwork or questionnaire exercise. The following sections describe a number of points which need to be borne in mind when conducting questionnaire studies. What follows only touches upon the complex issues involved in questionnaire exercises and those interested in pursuing the topic should consult the invaluable book by Dixon and Leach (1977a) and the classic text of Moser and Kalton (1971).

7.8.1 Sampling

The starting point in any questionnaire study is a consideration of the amount of required information. Assume that our concern is to identify the characteristics of households who have moved within the city in the last 18 months. The total number of such households is termed the ***population***. In an ideal situation the whole population would be interviewed. However, constraints of time and money, for most studies, impose an upper limit to the number of households who could be interviewed in a reasonable space of time. In most cases then only a ***sample*** can be interviewed. The correct sampling procedure is the one which allows us to make statements concerning the characteristics of the population on the basis of an analysis of the sample. A detailed consideration of sampling methods and procedures is given by Dixon and Leach (1977b). The larger the sample, the more accurate the sample and in general a sample of 30 is the minimum size from which to draw conclusions concerning the characteristics of the population.

7.8.2 Questionnaire techniques

Having identified the required sample size the next issue which arises is the proper method of obtaining the desired information. This issue can be subdivided into the inter-related parts — the boosting of response rates, the type of the questions and the layout of the questionnaire.

Response rates

Apart from the electoral register and the Census, people and households do not have to answer questions put to them. It is important, therefore, to arrange the questionnaire such that, as many of your sample as possible reply to your questions; this is termed boosting the response rates. It has been found that response rates are greatest when people are given some warning of your intention to interview them. Simply asking questions on the street of passers by is a frustrating and chastening experience. Besides it is very difficult to control your sampling in such circumstances. The greatest response rates are achieved when an initial letter is sent to the persons or households. The letter should be phrased such that the respondant knows why the study is being conducted and that the information they give will be held in the strictest confidence. ***Figure 7.5*** is a copy of a letter sent to recent movers in Bristol indicating that an interviewer would be calling upon them in a few days while ***Figure 7.6*** is a copy of a letter sent with a postal questionnaire to a sample of landlords in Bristol. The sample in the latter case was partly taken from the information available in the rent register (***see*** Section 7.4).

Response rates are greater when interviewers call upon respondants to administer the questionnaire rather than postal exercises in which respondants are sent a questionnaire and told to complete and return it. A useful compromise — given that administered exercises can be costly in the delivery and collection method — sometimes known as the drop and collect method, is one in which questionnaires are delivered with a note explaining that someone will call round in a few days to collect the completed questionnaire. This method generally gives higher response rates than the simple postal method.

Type of questions

The format of questions in a questionnaire should be such that they elicit the required response. Questions should be phrased in a clear and concise manner. At all times try to

UNIVERSITY OF BRISTOL

TELEPHONE 24161

DEPARTMENT OF GEOGRAPHY
UNIVERSITY ROAD
BRISTOL
BS8 1SS

Dear Householder,

I am conducting a survey into the movement patterns of people in Bristol. In the next few days an interviewer will call on you to ask a few questions about your past and present dwelling.

Your co-operation would be greatly appreciated and you may be assured that all information will be held in the strictest confidence and individual names will not be used. The interview will take between 15–20 minutes to complete.

If you have any queries or points to raise please feel free to write to me at the above address or telephone Bristol 24161, extension 692.

The information obtained is to be used for personal research for the degree of Ph.D. and is not an official University survey.

Yours sincerely,

John R. Short, M.A.

Figure 7.5 Introductory letter to households

University of Bristol

Department of Geography,
University Road,
Bristol,
BS8 1SS

Tel. 24161, Ext. 264.

Dear Sir or Madam,

I am currently engaged on a research project which is examining trends in the private rented sector of the Bristol housing market. This research is funded by the Social Science Research Council and the final report will hopefully have some effect on public appreciation of the problems facing landlords.

I have sent out a number of questionnaires to people I believe provide rented accommodation in Bristol. If you do provide such accommodation, I would be very grateful if you could complete and return the enclosed questionnaire in the prepaid envelope. It should only take ten to fifteen minutes, at the most, to complete. You may be assured that all responses will be treated in the strictest confidence. As you will see, no names are asked for or, indeed, needed, since the aim of this study is to collect information on landlords in general, not to identify individuals.

If you cannot provide all the information, please do not hesitate to return a partially completed form. You do not have to complete the questionnaire but your response will provide the basis for a more reasoned examination of the private rented sector than the ill-informed views prevalent at this time.

If you have any questions or points to raise, please do not hesitate to contact me at the above address. Copies of the final report will be deposited with the Bristol branch of the Property Owners Association.

Yours faithfully,

(Dr.) John R. Short

Encl:

Figure 7.6 Introductory letter to landlords

TABLE 7.6 INCOME AND AGE CATEGORIES

A. Income categories	B. Age categories
<£1000	<20
£1000–£2000	+20–30
+£2000–£3000	+30–40
+£3000–£5000	+40–50
+£5000–£10 000	+50–60
+£10 000	+60–65
	+65

keep the questions as short and as simple as possible. Questions which require a good memory on the part of the respondant, e.g. asking people where they shopped two months ago, are prone to error since memory recall may vary enormously in the sample. The question should be worded carefully so that respondants are not embarrassed or upset, since this may result in incorrect replies and, at worst, the termination of the interview. One student who went up to people in the street and asked them what social class they were in was surprised by some of the replies and the nature of the response!

In relation to age, income and other confidential items of information it is useful to ask respondants to give their replies in terms of broad categories. For example, rather than ask people precisely how much they earn after tax or how old they are, it is better to ask them to place their yearly earnings after tax and age in a broad category such as those shown in ***Table 7.6***. The categories should be broad enough so that respondants are not dissuaded from answering the question, but small enough for meaningful analysis.

Layout and structure

The public has a limited tolerance for answering questions and, unless there are very exceptional circumstances, questionnaires should not take more than 10–15 min to complete.

It is important that the early questions should be reasonably simple so that the respondant is put at ease at the beginning of the interview; more probing questions should be left until the end of the interview, by which time some degree of rapport should have been established between the interviewer and the respondant.

The questionnaire should be structured to broadly reflect the sequence of the decision-making process which is under consideration because this aids the memory recall of the respondants. ***Table 7.7***, for example, lists the character and order of questions to be asked in a study of residential mobility. Note how the order of questions broadly corresponds to the sequence of events found in the process of residential relocation.

Whatever the precise layout and type of question you decide to use, it is essential to conduct a pilot study in which the questionnaire is administered to a small sample. Such a trial run enables inconsistencies to be noted and possible sources of misunderstanding and confusion to be eliminated in subsequent versions of the questionnaire. You may even

TABLE 7.7 SEQUENCE OF QUESTIONS IN STUDY OF RESIDENTIAL MOBILITY

Type of question		Examples
1.	Past location	Location of previous address, price? number of rooms?
2.	Decision to seek new dwelling	What made you leave previous place?
3.	Search for vacancies	How many vacancies did you look at?
4.	Choice of specific dwelling	What made you choose this place?
5.	Present accommodation	Price? Number of rooms?
6.	Future mobility intentions	Do you intend to move within next 2 years?
7.	Background information	Age? Income?

find it necessary to conduct a series of pilot studies to ensure that the final version of your questionnaire contains no errors and elicits the desired information.

Questionnaire exercises should be treated with great caution. The public has limited tolerance and restricted time for filling in questionnaires and answering questions. There is always the danger that a series of questionnaire studies in one particular area may exceed this tolerance level and hence pre-empt researchers from undertaking more substantive pieces of research in the same area. Questionnaire exercises also need to be treated with caution since they require an adequate sampling frame, a proper sample, carefully worded questions and an ability to administer questionnaires. A failing in any of these departments can render the whole project invalid.

APPENDIX A POPULATION REDISTRIBUTION: AN EXERCISE

In *Tables A.1* and *A.2* are given the population totals for areas in Aberdeenshire and the Bristol region for 1951 and 1971. These data have been taken from the respective County Reports. *Figures A.1* and *A.2* show the boundaries of the constituent areas and the

TABLE A.1 POPULATION CHANGE IN ABERDEENSHIRE, 1951–1971

		Population in 1951	Population in 1971
	City of Aberdeen	182 729	182 071
Towns	Ballater	1032	982
	Kintore	870	835
	Huntly	4198	3645
	Turriff	2994	2858
Areas	Aberdeen	25 761	36 130
	Alford	9744	7011
	Deer	23 406	19 219
	Deeside	8897	6642
	Ellon	11 674	8505
	Garioch	8947	6863
	Huntly	5999	4314
	Turriff	8911	6210

TABLE A.2 POPULATION CHANGE IN THE BRISTOL REGION, 1951–1971

	Population in 1951	Population in 1971
Bristol	443 363	426 657
Axbridge	26 544	39 436
Clutton	18 048	20 845
Keynsham	8280	19 018
Kingswood	18 806	30 290
Long Ashton	22 975	42 111
Mangotsfield	17 121	23 284
Sodbury	38 393	67 124
Thornbury	25 830	43 094
Warmley	10 384	24 234

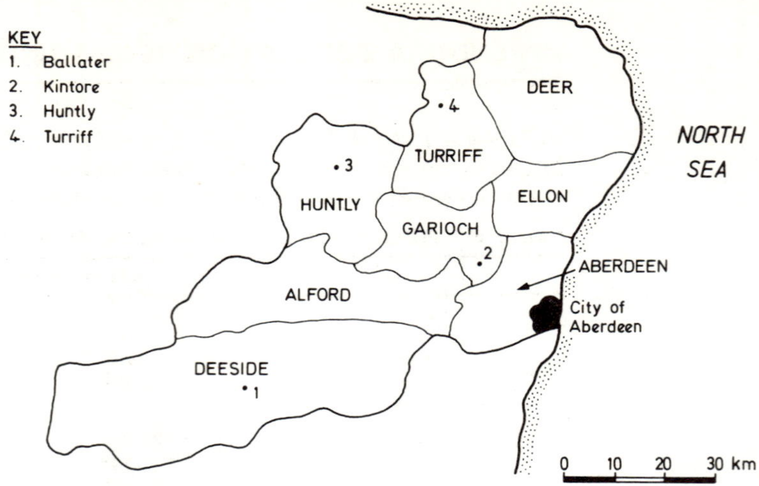

Figure A.1 A map of Aberdeenshire

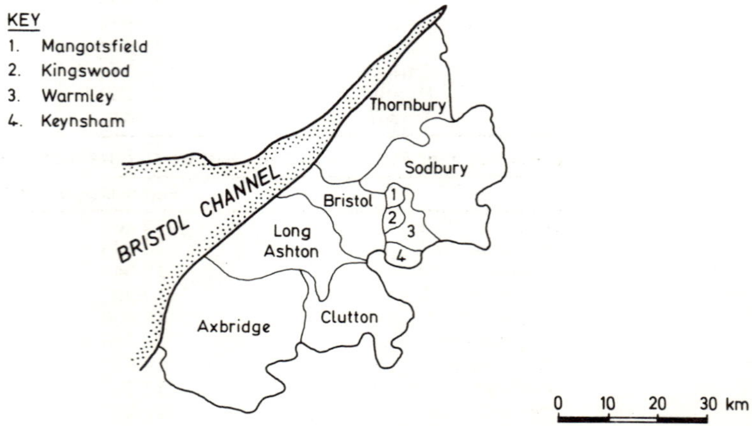

Figure A.2 A map of the Bristol region

location of small towns which make up the two urban regions. These data can be used as the basis for a three-part exercise which highlights the different types of population redistribution that can occur within different types of urban region.

A. (1) Calculate the percentage change* in population size for the different districts, small town and the city of Aberdeen in Aberdeenshire.

(2) Divide the ***districts*** into categories on the basis of population change and show the distribution of these categories on the map. Comment on the spatial pattern of population change.

(3) Which forms of population movement would you associate with the patterns you have described? Remember that Aberdeenshire is primarily an agricultural region. What are the 'push' and 'pull' factors behind these forms of movement?

(4) Comment on the patterns of population change in the small towns and the city of Aberdeen. What is the overall direction of population movement in this urban hierarchy? What are the likely consequences of these movements on services in the small towns?

B. (1) Calculate the percentage change in population size for the city of Bristol and surrounding districts.

(2) Divide the districts into categories on the basis of population change and show the distribution of these categories on the map. Comment on the spatial distribution.

*The percentage change is calculated by the following formula:

$$\frac{\text{1971 population} - \text{1951 population}}{\text{1951 population}} \times 100$$

Thus, the city of Bristol with a population of 443 363 in 1951 and 426 657 in 1971 experienced a population change of:

$$\frac{426\,657 - 443\,363}{443\,363} \times 100 = -3.76\%$$

(3) Which forms of population movement would you associate with the patterns you have described? Identify the 'push' and 'pull' factors behind this form of movement.

C. (1) Compare and contrast the patterns of population change in the two urban systems. Comment on the difference in scale.

(2) Does the use of percentage figures mask differences in absolute levels of change?

(3) Comment on the disadvantages of using differences in population totals in an analysis of population redistribution.

APPENDIX B CENSUS DATA: BRISTOL WARDS

TABLE B.1 SELECTED VARIABLES FOR BRISTOL WARDS FROM 1961 AND 1971 CENSUSES

Ward		Population in 1961	Population in 1971	Percentage private households in owner-occupation	Percentage private households renting from local authority	Percentage private households living at greater than 1.0 person/room	Percentage private households lacking exclusive use of hot water, fixed bath and inside W.C.
Avon	1	20 145	18 592	31.2	61.9	6.9	5.5
Bedminster	2	12 635	11 632	72.1	16.3	2.4	21.1
Bishopston	3	13 740	12 950	72.0	0.7	2.6	25.1
Bishopsworth	4	27 557	30 849	20.9	75.2	11.3	3.0
Brislington	5	17 231	17 242	64.0	25.9	3.3	11.7
Cabot	6	13 352	10 753	25.2	18.9	5.0	38.1
Clifton	7	14 892	14 695	22.0	1.0	4.4	35.7
District	8	15 325	14 579	48.2	2.1	9.1	42.6
Durdham	9	14 574	14 437	66.3	4.1	1.7	18.8
Easton	10	12 657	10 146	50.5	32.2	7.7	33.7
Eastville	11	16 177	14 674	67.0	20.0	4.6	22.1
Henbury	12	20 562	20 801	25.1	69.6	5.4	2.4
Hengrove	13	20 964	21 185	31.2	64.7	11.8	12.9
Hillfields	14	15 149	14 260	51.9	41.6	3.6	11.1
Horfield	15	15 584	14 691	45.8	48.1	6.5	7.9
Knowle	16	14 893	13 746	70.7	9.5	3.9	20.8
Redland	17	13 981	13 961	29.7	1.1	5.2	39.3
St. George East	18	14 360	16 563	72.2	20.1	3.2	15.0
St. George West	19	11 413	9800	64.4	21.6	3.0	32.5
St. Paul	20	13 400	8720	51.3	21.5	11.5	43.5
St. Philip and Jacob	21	9001	5159	13.6	69.6	4.7	18.9
Somerset	22	15 923	14 628	40.3	54.8	6.6	9.7
Southmead	23	20 415	19 085	28.4	66.3	9.2	8.0
Southville	24	10 393	9362	62.3	12.8	3.4	32.2
Stapleton	25	19 312	18 263	59.9	28.8	2.0	12.0
Stockwood	26	16 395	27 816	56.3	35.3	3.7	1.9
Westbury	27	16 490	18 460	81.8	5.2	0.7	4.7
Windmill Hill	28	10 528	9180	65.7	14.3	7.0	37.3
Bristol County Borough		437 048	426 229	48.9	31.3	5.4	18.2

Percentage population born N.C.	Percentage population aged 0–14 years	Percentage population aged 15–34 years	Percentage population aged >60	Percentage persons in socio-economic groups 1,2,3,4,13	Percentage persons in socio-economic groups 5,6	Percentage persons in socio-economic groups 8,9,12,14,15	In-migrants (1966–1971)
0.4	26.5	27.0	14.4	7.0	10.7	43.1	3.7
0.3	17.7	22.5	28.1	9.3	19.3	42.9	2.9
2.6	16.7	28.5	23.6	17.1	29.6	35.4	13.1
0.5	32.7	29.3	9.5	7.1	12.7	46.9	1.2
0.4	21.9	25.7	20.7	10.3	15.8	44.9	2.1
1.9	13.3	37.0	23.1	20.9	21.3	28.2	5.8
3.2	12.2	44.5	21.0	37.0	32.4	16.7	31.1
11.5	21.0	32.1	19.9	14.4	20.5	30.0	4.6
1.3	17.9	27.0	24.8	39.0	35.3	16.9	19.4
10.3	22.5	22.6	25.4	4.0	8.5	48.9	4.2
4.0	21.9	24.7	21.8	10.1	19.8	44.6	5.5
0.7	23.6	27.4	15.5	20.7	23.3	31.3	6.1
0.6	28.8	27.2	16.3	8.0	13.3	40.1	3.3
0.6	21.6	22.8	25.1	8.8	15.5	50.7	4.6
1.1	23.3	25.2	21.1	9.4	20.4	42.0	5.6
2.0	20.4	25.7	25.4	13.7	24.8	37.5	7.2
2.3	12.6	43.5	21.0	24.7	33.5	28.1	27.8
0.4	22.9	26.2	21.5	12.6	19.6	45.5	6.6
1.3	17.5	22.8	29.8	7.1	13.6	44.3	4.9
13.0	24.7	23.6	22.5	4.7	10.3	38.8	4.2
1.2	12.0	19.3	37.4	8.9	15.2	51.5	12.1
0.5	23.1	24.9	20.1	9.3	14.1	43.9	2.8
1.3	25.0	27.9	17.4	14.0	18.6	36.9	2.2
0.9	18.6	25.5	25.3	11.4	14.7	43.3	8.0
0.6	15.1	22.7	28.4	17.7	22.2	41.4	9.5
0.5	27.8	28.8	13.2	18.2	24.3	37.9	13.4
1.6	19.8	28.0	20.2	55.3	25.8	8.6	15.9
4.0	21.4	26.1	22.9	10.4	14.7	44.1	17.2
2.0	22.0	27.9	20.4	15.3	19.7	38.6	8.7

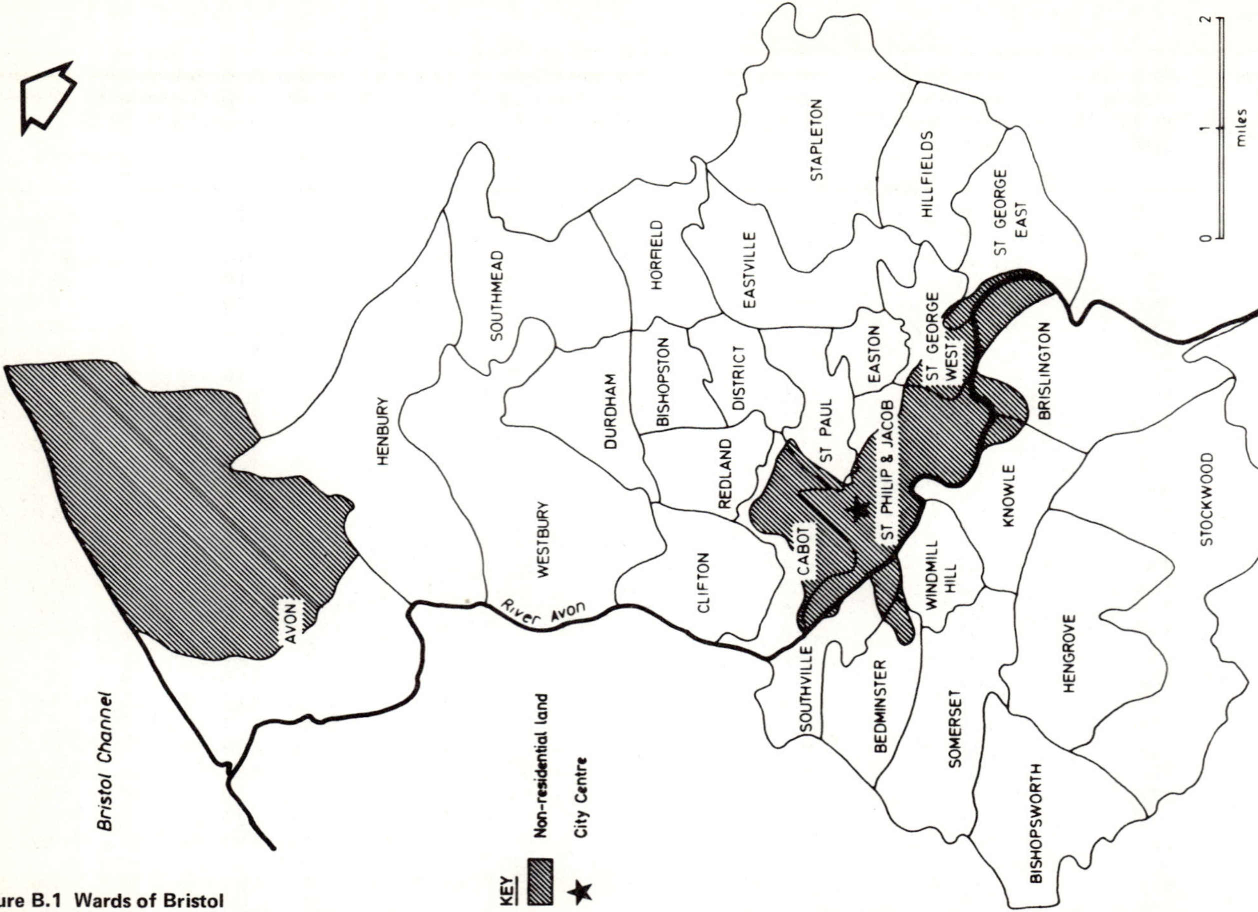

Figure B.1 Wards of Bristol

REFERENCES

Abler, R., Adams, J.S. and Gould, P. (1971). ***Spatial Organization: The Geographer's View of the World.*** Englewood Cliffs, N.J.; Prentice Hall

Ambrose, P. and Colenutt, B. (1975). ***The Property Machine***. Harmondsworth; Penguin

Bassett, K. and Short, J.R. (1978). 'Housing improvement in the inner city.' ***Urban Studies*** **15**, 333–342

Berry, B.J.L. (1961). 'City size distributions and economic development.' ***Econ. Develop. Cult. Change*** **9**, 573–588

Berry, B.J.L. (1967). ***Geography of Market Centres and Retail Distribution***. Englewood Cliffs, N.J.; Prentice Hall

Boaden, N. (1971). ***Urban Policy Making.*** Cambridge; Cambridge University Press

Brush, J.E. and Bracey, H.E. (1955). 'Rural service centres in northern England.' ***Geogrl Rev.*** **45**, 559–569

Bull, C.J. and Bull, P.J. (1978). 'Regional directories as a potential source for the study of intra-urban manufacturing industry.' ***Geography*** **63**, 198–204

Burgess, E.W. (1925). 'The growth of the city.' In ***The City***. Ed. R.E. Park *et al*. Chicago; University of Chicago Press

Carter, H. (1972). ***The Study of Urban Geography***. London; Edward Arnold

Cordey-Hayes, M. and Gleave, D. (1973). 'Migration movements and the differential growth of city regions in England & Wales.' ***Centre for Environmental Studies Research Paper 1***

Daultrey, S. (1976). ***Principal Components Analysis.*** Catmog 8. Geo Abstracts. Norwich

Dixon, C. and Leach, B. (1977a). ***Questionnaire and Interviews in Geographical Research***. Catmog 18. Geo Abstracts. Norwich

Dixon, C. and Leach, B. (1977b). ***Sampling Methods for Geographical Research***. Catmog 17. Geo Abstracts. Norwich

Duncan, O.D. and Duncan, B. (1955). 'Residential distribution and occupational stratification.' ***Am. J. Sociol.*** **60**, 493–506

Edwards, B. (1972). ***Sources of Economic and Business Statistics***. London; Heinemann

Edwards, B. (1974). ***Sources of Social Statistics***. London; Heinemann

Everson, J.A. and Fitzgerald, B.P. (1972). ***Inside the City***. London; Longman

Fielding, A.J. (1971). 'Internal migration in England and Wales.' ***Centre for Environmental Studies*** UWP 14

Guy, C. (1976). 'The location of shops in the Reading area.' ***University of Reading Geographical Papers***. Number 46

Hall, P., Gracey, H., Drewett, R. and Thomas, R. (1973). ***The Containment of Urban England***. London; P.E.P. and Allen & Unwin

Hamilton, F.E.I. (1967). 'Models of industrial location.' In ***Models in Geography***. Ed. R.J. Chorley and P. Haggett. London; Methuen

Herbert, D.T. (1972). ***Urban Geography: A Social Perspective***. Newton Abbot; David & Charles

Herbert, D.T. (1973). 'Residential mobility and preference: a study of Swansea.' In ***Social Patterns in Cities***. Ed. B.D. Clark and M.B. Gleave. Institute of British Geographers Special Publication No. 5

Horwood, E.M. and Boyce, R.R. (1959). ***Studies of the Central Business District and Urban Freeway Development***. Seattle; University of Washington Press

House, J.W. (1953). 'Medium sized towns in the urban pattern of two industrial societies: England and Wales – U.S.A.' ***Plann. Outl.*** **3**, 52–79

Hoyt, H. (1939). ***The Structure and Growth of Residential Neighbourhoods in American Cities***. Washington, D.C.; Federal Housing Administration

Johnston, R.J. (1971). ***Urban Residential Patterns***. London; Bell

Johnston, R.J. (1976). 'Residential area characteristics: research methods for identifying urban sub-areas-social area analysis and factorial ecology.' In ***Social Areas in Cities***, Vol. 1. Ed. D.T. Herbert and R.J. Johnston. London; Wiley

Johnston, R.J. (1978). ***Multivariate Statistical Analysis in Geography***. London; Longman

Jones, T.P. and McEvoy, D. (1978). 'Race and space in cloud-cuckoo land.' ***Area*** **10**, 162–166

Kivell, P.T. (1972). 'A note on metropolitan areas 1961–1971.' ***Area*** **4**, 179–184

Lee, T.R. (1973). 'Ethnic and social class factors in residential segregation: some implications for dispersal.' ***Envir. & Plann.*** **5**, 477–490

Lever, N.F. (1972). 'The intra-urban movement of manufacturing: a Markov approach.' ***Trans. Inst. Br. Geogr.*** **56**, 21–38

Masser, I. (1970). 'A test of some models for predicting intermetropolitan movement of population in England and Wales.' ***Centre for Environmental Studies***, UWP 9

Moser, C. and Kalton, G. (1971). ***Survey Methods in Social Investigation***. London; Heinemann

Murdie, R.A. (1976). 'Spatial form in the residential mosaic.' In ***Social Areas in Cities,*** Vol. 1. Ed. D.T. Herbert and R.J. Johnston. London; Wiley

Robinson, N.S. (1950). 'Ecological correlations and the behaviour of individuals.' ***Am. sociol. Rev.*** **XV**, 351–357

Robson, B.T. (1973). ***Urban Growth: An Approach***. London; Methuen

Robson, B.T. (1975). ***Urban Social Areas***. London; Oxford University Press

Rowley, G. and Shepherd, P.M. (1976). 'A source of elementary spatial data for town centre research in Britain.' ***Area*** **8**, 201–208

Short, J.R. (1976). 'Aspects of Residential Mobility in Bristol.' ***PhD Thesis, University of Bristol***

Short, J.R. (1978a). 'Residential mobility in the private housing market of Bristol.' ***Trans. Inst. Br. Geogr.,*** New Series **3**, 533–547

Short, J.R. (1978b). 'Population turnover: problems in analysis and an alternative method.' ***Area*** **10**, 231–235

Smith, D.M. (1977). ***Human Geography: A Welfare Approach***. London; Edward Arnold

Wise, M. (1949). 'On the evolution of the jewellery and gun quarters in Birmingham.' ***Trans. Inst. Br. Geogr.*** **15**, 57–72

Woods, R.I. (1976). 'Aspects of the scale problem in the calculation of segregation indices: London and Birmingham, 1961 and 1971.' ***Tidsskr. Econ. Soc. Geogr.*** **67**, 169–174

INDEX